Combinatorial Libraries
Synthesis, Screening and Application Potential

Combinatorial Libraries

Synthesis, Screening and Application Potential

Editor
Riccardo Cortese

Walter de Gruyter · Berlin · New York 1996

Editor
Professor Dr. Riccardo Cortese
Scientific Director
IRBM-Istituto di Ricerche di Biologia Molecolare P. Angeletti
Via Pontina Km. 30,600
00040 Pomezia (Roma)

Italy

With 92 figures

Cover illustration
Molecular-graphical representation of the zinc-finger peptide library. The main chain of the 26-residue zinc finger peptide is shown as a pink tube. The selectable positions O_1-O_2-O_3-O_4-O_5 are shown as yellow spheres centered on the side-chain β-carbon; the zinc atom and the coordinating cysteine and histidine residues are in white. Further details can be found in Bianchi et al., *Journal of Molecular Biology* 247, 154–160 (1995) Academic Press. Graphics courtesy of Anna Tramontano, IRBM; © IRBM. P. Angeletti, 1995.

Library of Congress Cataloging-in-Publication Data

Combinatorial libraries : synthesis, screening, and application potential / editor, Riccardo Cortese.
Includes index.
ISBN 3-11-014395-X
1. Nucleotide sequence. 2. Amino acid sequence. 3. Genetic vectors. I. Cortese, Riccardo.
QP620.C64 1995
574087´328 – dc20 95–39772
CIP

Die Deutsche Bibliothek – Cataloging-in-Publication Data

Combinatorial libraries : synthesis, screening and application potential / ed. Riccardo Cortese. – Berlin ; New York : de Gruyter, 1995
ISBN 3-11-014395-X
NE: Cortese, Riccardo [Hrsg.]

⊗ Printed on acid-free paper which falls within the guidelines of the ANSI to ensure permanence and durability.

Converting and typesetting by: Frohberg GmbH, Freigericht. – Printing: Karl Gerike GmbH, Berlin. – Binding: Dieter Mikolei, Berlin. – Cover Design: Hansbernd Lindemann, Berlin.
Printed in Germany.

Introduction

The number of orphan receptors and antibodies may diminish in the future thanks to the emergence of novel experimental strategies which could facilitate the identification and characterization of their respective ligands.

The common underlying principle of these techniques is based on the generation, in vitro or in vivo, of expansive collections (libraries) of very large numbers of molecules (peptides, oligonucleotides or other) or randomized sequences or structures, one or several of which may possess the desired characteristics. It is clear that the higher the diversity of the library, the higher the chance of hitting the target. However, although the theoretical multiplicity of these libraries is almost astronomical, practical aspects impose strict limitations on the numbers of molecules that can be handled per experiment. A great deal of ingenuity thus went into converting this, undoubtedly, great idea into a facile, widely-used, user-friendly technique. As a result, developments of "combinatorial" synthetic procedures helped overcome the difficulties linked with the construction of the libraries, while the setting-up of highly-powerful "screening" and "selection" methods facilitated the identification of the "hits" with the desired properties. Considering the multiplicity of the different possible structures, this could be likened to presenting the person, trying to find the proverbial needle in a haystack, with a magnet.

This revolutionary technology is still at an early stage of development and thus the relative merits of the different approaches can be assessed mostly at a theoretical level. For this reason, the articles presented in this compilation cover a broad spectrum of ideas and applications. However, there are already numerous indications which convince us that the concept of combinatorial libraries has opened up new frontiers of basic and applied biology and chemistry.

Contents

C Phage Display of Peptide Libraries

6 Structural and Functional Constraints in the Display of Peptides on Filamentous Phage Capsids

Gianni Cesareni, Olga Minenkova, Luciana Dente, Gioacchino Iannolo, Adriana Zucconi, Manuela Helmer Citterich, Alessandra Lanfrancotti, Luisa Castagnoli and Costantino Vetriani

7 Conformationally Defined Peptide Libraries on Phage: Selectable Templates for the Design of Pharmacological Agents

Maurizio Sollazzo, Elisabetta Bianchi, Franco Felici, Riccardo Cortese, Antonello Pessi

8 Discovery of Disease-Specific Mimotopes by Screening Phage Libraries with Human Serum Samples

Alfredo Nicosia, Paolo Monaci, Allessandra Luzzago, Giovanni Galfré, Franco Felici, Caterina Prezzi, Carmela Mennuni, Annalisa Meola, Monica Mecchia and Riccardo Cortese

D Phage Display of Protein Domains

11 Isolating High Affinity Human Antibodies from Phage Repertoires

Kevin FitzGerald, David Chiswell, John Earnshaw, Rodger Smith, John Kenten, Richard Williams and John McCafferty

12 Altering the Function of Enzymes and Macromolecular Inhibitors by Phage Display

Qing Yang, Cheng-I Wang and Charles S. Craik

A Synthetic Peptide Libraries

1 Soluble Synthetic Combinatorial Libraries: The Use of Molecular Diversities for Drug Discovery

Barbara Dörner, Sylvie E. Blondelle, Clemencia Pinilla, Jon Appel, Colette T. Dooley, Jutta Eichler, John M. Ostresh, Enrique Pérez-Payá and Richard A. Houghten

1.1 Introduction

Methods for the generation and systematic screening of immense molecular diversities (i.e., tens to hundreds of millions of compounds) have recently been developed and represent a powerful tool in the search for novel pharmacological agents [reviewed in (Gallop et al., 1994; Gordon et al., 1994; Blondelle et al., 1995a; Pinilla et al., 1995)]. A number of different strategies based on the principle of solid phase synthesis are now available to prepare these diversities in a manner which permits their ready use in biological screening. The synthesis of libraries was originally focused on peptides and nucleotides, for which synthetic procedures were straightforward and well established. Peptides are key mediators of biochemical information and have long been used as starting compounds for the development of novel drugs, even though they have limitations as potential therapeutic agents due to their typical lack of oral activity, susceptibility to proteolytic breakdown, and inability to pass through the blood brain barrier. Due to these limitations, recent trends in this field have been directed toward the preparation of chemical libraries as potential sources for improved leads for drug discovery. Powerful chemical methods have been developed which take advantage of solid phase synthetic procedures for the generation of a wide range of compounds having novel physiological properties. Thus, a variety of novel solid phase-based chemistries enable the generation of libraries of organic molecules in an efficient and reproducible manner.

This chapter briefly reviews a number of approaches developed by and used in our laboratory to generate soluble synthetic combinatorial libraries of peptides, peptidomimetics and organic compounds, and illustrates their use for basic research and drug discovery.

1.2 Synthetic Combinatorial Libraries (SCLs)

The original definition of the library concept as applied to peptides referred to large numbers (up to trillions) of compounds prepared in a highly systematic manner, which consis- ted of all combinations of subunits of a class of compounds termed combinatorial libraries (Geysen et al., 1986; Houghten et al., 1991; Lam et al., 1991; Pinilla et al., 1994b). Peptidomimetic or organic combinatorial libraries have been generated in a range of thousands to millions of compounds (Zuckermann et al., 1994; Ostresh et al., 1994a; Cuervo et al., 1995). The term "-library" is now commonly used to describe collections of compounds or series of analogs ranging in number from less than 10 to hundreds (Fodor et al., 1991; Cho et al., 1993; Bunin et al., 1994; DeWitt et al., 1993). The premise of such libraries is that they enable completely novel, biologically active compounds to be identified through screening without any prior structural or sequence knowledge.

The peptide library approaches presented to date fall into three broad categories, the difference being the manner in which the sequences are synthesized and/or screened. The first category represents synthetic approaches, in which peptide mixtures are synthesized, cleaved from their solid support and assayed as free compounds in solution. This category includes synthetic combinatorial libraries (SCLs) of peptides developed in this laboratory and elsewhere (Houghten et al., 1991; Owens et al., 1991; Hortin et al., 1992; Blake and Litzidavis, 1992; Simon et al., 1992). The second category includes synthetic approaches, in which peptide mixtures are synthesized and assayed while attached to either plastic pins (Geysen et al., 1986; Geysen and Mason, 1993), resin beads (Lam et al., 1991; Lam and Lebl, 1992), or cotton (Eichler et al., 1994a). A modification of this approach uses encoded "tags" to allow the rapid identification of peptides that cannot be sequenced directly from individual beads (Needels et al., 1993; Kerr et al., 1993; Nikolaiev et al., 1993; Ohlmeyer et al., 1993). The third category includes the molecular biology approaches, in which peptides or proteins are presented on the surface of filamentous phage particles or plasmids. This was-first presented in 1990 (Scott and Smith, 1990; Devlin et al., 1990; Cwirla et al., 1990) and was recently reviewed (Smith et al., 1993; Scott and Craig, 1994).

In 1991 (Houghten et al., 1991), this laboratory first presented a variety of peptide SCLs that are soluble and therefore able to freely interact with acceptor systems, and can be used in virtually any bioassay system [reviewed in (Blondelle et al., 1995a; Pinilla et al., 1994a; Houghten, 1994)]. These libraries were prepared using the simultaneous multiple peptide synthesis method [SMPS – also known as the T-bag approach (Houghten, 1985)]. Each resulting soluble peptide mixture within a library is characterized by single or multiple defined residue(s) at given position(s). No decoding is necessary since mixtures are categorized in separate containers and therefore readily identifiable. The iterative process of selection and synthesis (Houghten et al., 1991) and the positional scanning SCLs (described below; Pinilla et al., 1992) are the strategies (or deconvolution processes) devised for the systematic identification of active sequences from mixtures of millions of compounds present in a combinatorial library.

1.2.1 Iterative Process

The first two soluble SCLs prepared (Houghten et al., 1991; Houghten et al., 1992) consisted of six-residue peptide sequences having either acetylated (Ac) or non-acetylated N-termini and amidated C-termini. The first two amino acids in each peptide chain were individually and specifically defined, while the last four amino acids consisted of approximately equimolar mixtures of 19 of the 20 natural L-amino acids. Cysteine was omitted from the mixture positions of the two SCLs, but included in the defined positions. These libraries can be represented by the general formulas Ac-O_1O_2XXXX-NH_2 and O_1O_2XXXX NH_2, with O_1 and O_2 equal to AA, AC, AD, etc. through YV, YW, YY, for a total of 400 combinations

Table 1.1: Iterative process illustrating the screening of a library in which both the O and X positions consist of the 20 L-amino acids

Step	Process	Sequence	Group		# of peptides/group	Total # of peptides
1.	Screening	OOXXXX	400	x	160,000	64,000,000
	Selection	RRXXXX	1	x	160,000	160,000
2.	Synthesis	RROXXX	20	x	8,000	160,000
3.	Screening	RROXXX	20	x	8,000	160,000
	Selection	RRWXXX	1	x	8,000	8,000
4.	Synthesis	RRWOXX	20	x	400	8,000
5.	Screening	RRWOXX	20	x	400	8,000
	Selection	RRWCXX	1	x	400	400
6.	Synthesis	RRWCOX	20	x	20	400
7.	Screening	RRWCOX	20	x	20	400
	Selection	RRWCKX	1	x	20	20
8.	Synthesis	RRWCKO	20	x	1	20
9.	Screening	RRWCKO	20	x	1	20
	Selection	RRWCKR	1	x	1	1

(20^2); each X position represents an equimolar mixture of the 19 amino acids. Four mixture positions result in a total of 130,321 combinations (19^4). Each of the 400 different peptide mixtures which make up a library thus consists of 130,321 individual hexamers, which in total represent 52,128,400 peptides per SCL.

The most effective peptide mixture(s) are selected in the initial direct screening of the separate mixtures making up the peptide library. An iterative process is then carried out in which the subsequent X positions of the peptide mixtures are individually defined with each of the 20 natural L-amino acids as (Table 1.1; Blondelle et al., 1995a). At each iterative step, new soluble peptide mixtures having one additional defined position are assayed to identify the most active amino acid at the newly defined position. This process involves ranking, selecting and reducing the number of peptide sequences while synthetically defining one more position at each step. In some instances more than one peptide mixture is moved ahead, especially if chemically distinct amino acids (i.e., aspartic acid and lysine; phenylalanine and glycine) with similar activities are found upon screening the mixtures making up the peptide library.

Two hexapeptide SCLs composed entirely of D-amino acids have been generated in a similar manner, as well as a tetrapeptide SCL in which L-, D-, and unnatural amino acids were incorporated (Blondelle et al., 1994). The systematic grouping of the peptide mixtures, combined with their soluble character, allow the direct identification of the active peptide mixture (i.e., their inherent defined amino acids) from the nonactive ones without the extra step of sequencing or, as in the case of D- or unnatural amino acids, the drawbacks of incorporating a tag.

1.2.2 Positional Scanning SCLs

The positional scanning SCL (PS-SCL) approach enables active sequences to be identified in a single screening assay (Pinilla et al., 1994a; Pinilla et al., 1992; Dooley and Houghten, 1993). PS-SCLs are composed of individual positional SCLs in which a single position is defined with a single residue, while the remaining positions are composed of mixtures of residues (Table 1.2; Blondelle et al. 1995a). The defined position is "walked" through the entire sequence of the PS-SCL. Therefore, the number of positional SCLs is equal to the number of residues or building blocks in each compound of the PS-SCL. It should also be noted that each positional SCL, while addressing a single position of the sequence, represents the same collection of individual compounds. When used in concert, the data derived from each positional SCL yield information about the most important residues or building blocks for every position. The information can be used to synthesize individual compounds representing all possible combinations of the most active residues or building blocks at each position. The synthesis of these defined sequences confirms the information obtained from the initial screening of the separate positional libraries, and enables the identification of the most active compounds.To demonstrate this approach, the first PS-SCL, composed of L-amino acid hexapeptides (Pinilla et al., 1994a; Pinilla et al., 1992; Dooley and Houghten, 1993), is discussed. This library consists of six separate soluble positional SCLs, each composed of 20 different peptide mixtures having a single position defined with one of the 20 natural L-amino acids (represented as O), with the remaining five positions composed of mixtures of 18 L-amino acids (represented as X; cysteine and tryptophan omitted). The six positional SCLs differ only in the location of the defined position and can be represented as Ac-O_1XXXXX-NH_2, Ac-XO_2XXXX-NH_2, Ac-XXO_3XXX-NH_2, Ac-XXXO_4XX-NH_2, Ac-XXXXO_5X-NH_2, and Ac-XXXXXO_6-NH_2 (120 peptide mixtures in total). Each peptide mixture represents approximately 2.5 million (195) individual sequences; each of the six positional peptide libraries contains over 50 million hexamers. As for the previously described hexapeptide libraries, this PS-SCL was also prepared with a free N-terminus (Dooley and Houghten, 1993). In a similar manner, a soluble decapeptide PS-SCL was generated which represents a diversity of over 4 trillion decapeptides (Pinilla et al., 1994b). Two hexapeptide PS-SCL (one N-acetylated, the other one with free N-termini) have recently been prepared in which two positions are defined. These can be represented by Ac-O_1O_2XXXX-NH_2, Ac-XXO_3O_4XX-NH_2 and Ac-XXXXO_5O_6-NH_2. This PS-SCL is composed of 1,200 (3 x 400) peptide mixtures, each composed of 130,321 (19^4) hexapeptides.

Table 1.2: Positional scanning process illustrating the screening of a library in which both the O and X positions consist of the 20 L-amino acids

Step	Process	Sequence	Group		# of peptides/group	Total # of peptides
1.	Screening	OXXXXX	20	x	3,200,000	64,000,000
	Screening	XOXXXX	20	x	3,200,000	64,000,000
	Screening	XXOXXX	20	x	3,200,000	64,000,000
	Screening	XXXOXX	20	x	3,200,000	64,000,000
	Screening	XXXXOX	20	x	3,200,000	64,000,000
	Screening	XXXXXO	20	x	3,200,000	64,000,000
	Selection	RXXXXX	1	x	3,200,000	3,200,000
	Selection	XRXXXX	1	x	3,200,000	3,200,000
	Selection	XXWXXX	1	x	3,200,000	3,200,000
	Selection	XXXCXX	1	x	3,200,000	3,200,000
	Selection	XXXXKX	1	x	3,200,000	3,200,000
	Selection	XXXXXR	1	x	3,200,000	3,200,000
2.	Synthesis	RRWCKR	1	x	1	1

The advantages of the PS-SCL approach are that each position of the library is screened simultaneously, and the results found for each position enable the identification of individual, active compounds (peptides in the example shown) without the iterative synthesis and selection steps necessary when using single SCLs as described above. It should be noted that following the screening of a PS-SCL, an iterative process can always be initiated from any position of the sequence if necessary. Although often illustrated using peptide-based SCLs, these two strategies can also be applied to the identification of active peptidomimetic (Ostresh et al., 1994a) and organic compounds (Cuervo et al., 1995).

1.3 Synthetic Methods for the Generation of SCLs

1.3.1 Preparation of Combinatorial Peptide Mixtures

The peptide mixtures making up the SCLs, as well as the peptide mixtures synthesized for the iterative steps, were prepared using methylbenzhydrylamine (MBHA) polystyrene resin and standard tert.butyloxycarbonyl (Boc) chemistry in combination with simultaneous multiple peptide synthesis (SMPS) (Houghten, 1985). Peptide mixture resins were prepared either by a process termed divide, couple and recombine [DCR (Houghten et al., 1991)], or by using a mixture of amino acids that are incorporated simultaneously during the coupling procedure (Pinilla et al., 1992; Eichler and Houghten, 1993; Ostresh et al., 1994b). The side chain protecting groups of the peptide mixtures were removed and the peptides cleaved from their respective resins using the low-high hydrogen fluoride method (Houghten et al., 1986; Tam et al., 1983). The mixtures were then extracted with water or dilute acetic acid and the solutions lyophilized.

The DCR process (also known as the "split resin" method and the one bead/one peptide approach) was independently reported by this laboratory and two others in 1991 (Houghten et al., 1991; Lam et al., 1991; Furka et al., 1991). For the synthesis of libraries of peptides consisting of all combinations of the 20 naturally occurring L-amino acids, the coupled resin is divided into 20 equal portions, then each portion is separately coupled to a single amino acid. The 20 equal portions are mixed together for the required deprotection and neutralization steps. This resin mixture is then redivided again into 20 equal portions for the next individual coupling steps. This process is repeated until the synthesis is complete (Figure 1.1A). It should be noted that to provide high reproducibility during the full iterative process, portions of the resin are kept aside after each mixing step for the synthesis of the following iterative peptide mixtures. The essence of this approach is that highly complex mixtures of compounds can be prepared, each on a separate bead [i.e., millions of beads/millions of peptides (Lam et al., 1991)],

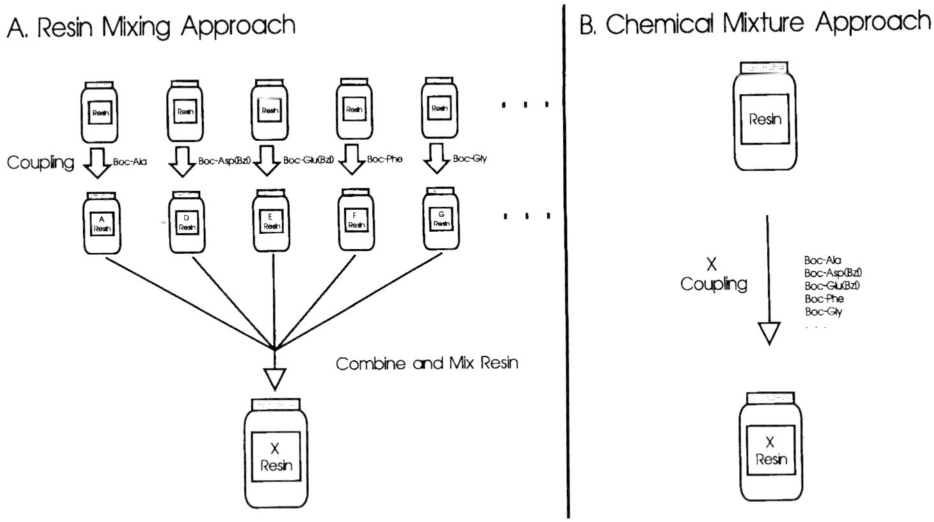

Figure 1.1: Generation of resin mixture using A) the divide, couple, and recombine process and B) chemical ratio of amino acids. A. For the preparation of peptides composed of all possible combinations of the 20 L-amino acids, the resins are divided and compartmentalized into 20 separate polypropylene packets at each coupling step, then mixed prior to the wash, deprotection and neutralization steps (Houghten et al., 1992). B. A mixture of amino acids of a predetermined ratio (Ostresh et al., 1994b) is used at each coupling step.

or, if the final pooled mixtures of resin-bound compounds are cleaved from the resin support, then virtually equal amounts of hundreds of thousands of compounds can be prepared in soluble form (Houghten et al., 1991). The DCR process, by its very nature (i.e., the exactitude of the physical process of weighing the separate resins, mixing them, subdividing and weighing them again), assures the generation of equimolar or very close to equimolar compound-bound resins. The DCR process also allows for the straightforward and direct incorporation of D- and unnatural amino acids into mixture or defined positions of a combinatorial library. The enlarged repertoire of amino acids in such libraries expands the chemical diversity that can be screened for biological activity. The DCR method, however, is limited in the number of compounds that can be synthesized due to the amount of resin required to include each compound in the library and the amount of resin required to achieve approximate equimolarity within the mixture. Thus, the production of approximately equimolar libraries of diversitiesgreater than 10^6 using the DCR method is impractical.

In a second synthetic approach, mixtures of all of the desired protected amino acids or building blocks are used simultaneously in the coupling step during solid phase synthesis (Figure 1.1 B) (Pinilla et al., 1992; Dooley and Houghten, 1993). If one uses the 20 naturally occurring L-amino acids, then six couplings of such a mixture will result in 20^6 hexamers. One perceived disadvantage of this method is the potential non equimolarity of the resultant mixtures due to the inherent variations in coupling rates between amino acids or building blocks. In the case of peptide libraries, kinetic studies of the relative coupling rates of incoming amino acids performed in this laboratory have led to the determination of the amino acid ratios necessary to ensure close to equimolar peptide mixtures within a library (Eichler and Houghten, 1993; Ostresh et al., 1994b). The success of this second approach has been reported extensively by this laboratory in comparative biological studies using peptide libraries representing the same peptide diversity but prepared by either method. Thus, identical sequences recognized by a monoclonal antibody raised against a peptide (Pinilla et al., 1994a; Pinilla et al., 1992), as well as peptide sequences having binding affinity to opioid receptors (Pinilla et al., 1994a; Dooley and Houghten, 1993), have been identified through the screening of two libraries prepared by the two approaches. The advantages of this method over the DCR method are the ease of the synthesis process, as well as the ability to prepare complete libraries with more than three or four mixture positions. Furthermore, the defined positions in the library can be readily located in positions other than the N-terminal.

1.3.2 Synthesis of a Cyclic Peptide Template Library

The character of linear peptides can easily be changed through cyclization. The reduced flexibility of cyclic peptides is thought to increase their potential for high affinity binding to various acceptor molecules. Furthermore, cyclic peptides have been reported to be resistant to enzymatic degradation (DiMaio and Schiller, 1980; Sham et al., 1988). Cyclic peptides have also been used as templates for the construction of conformationally defined molecules, as well as "surface mimetics" of discontinuous binding sites of proteins (Tuchscherer et al., 1993). The various methods for peptide cyclization on the solid support through lactam formation, which have also been used for the synthesis of cyclic peptide mixtures (Darlak et al., 1994), provided the methodological tools for the synthesis of cyclic peptide combinatorial libraries. A cyclic template combinatorial library recently prepared in the positional scanning format is composed of three positional libraries, which can be represented as *cyclo*[Lys(O_1)-Lys(X)-Lys(X)-Glu]-Gly-OH, *cyclo* [Lys(X)-Lys(O_2)-Lys(X)-Glu]-Gly-OH and *cyclo*[Lys(X)-Lys(X)-Lys(O_3)-Glu]-Gly-OH (Eichler et al., 1994b). The ε-amino groups of the three lysine residues were acylated using 10 carboxylic acids, representing structural elements such as heterocycles, xanthene, adamantane and norbornane, in addition to 19 of the 20 proteinogenic amino acids (cysteine was omitted in the mixture position). Thus, each

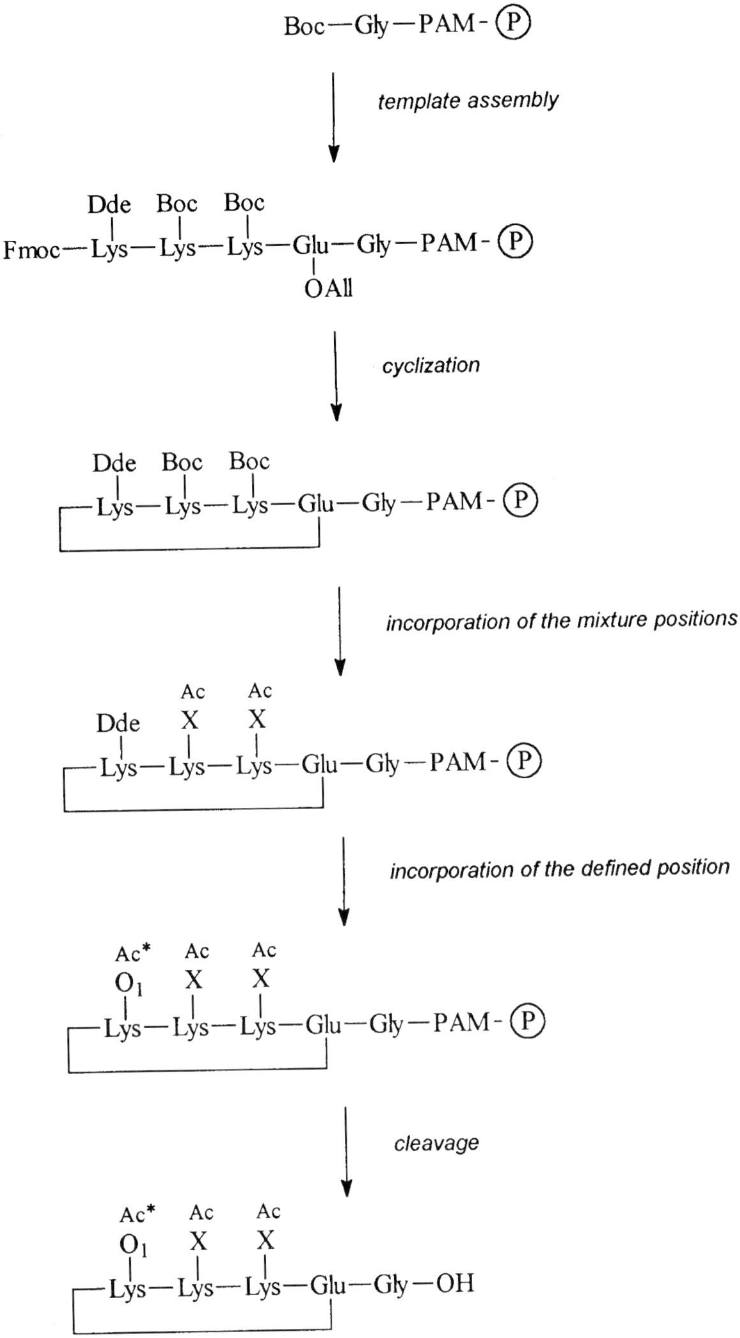

Figure 1.2: Preparation of a cyclic peptide template combinatorial library. Synthesis scheme for the first positional scanning library, cyclo[Lys(O_1)-Lys(X)-Lys(X)-Glu]-Gly-OH (Ac: Only acetylated if O is an amino acid) as described in (Eichler et al., 1994b).

positional library consisted of 30 separate peptide mixtures, with each mixture composed of 841 (29^2) individual compounds. Since solid phase chemistry was used [Boc-Gly-PAM resin (PAM: phenylacetamidomethyl) as solid support, 9-fluorenylmethoxycarbonyl (Fmoc) standard strategy for amino acid assembling], an important prerequisite for the synthesis (Figure 1.2) of such libraries was the availability of orthogonal protecting groups for the α- and ε-amino groups of lysine, as well as for the γ-carboxylic function of glutamic acid [Fmoc, Boc, Dde (Dde: (1-(4,4-dimethyl-2,6-dioxocyclohex-1-ylidene)ethyl), and OAll (OAll: allyl ester) were used]. Cyclization was performed between the terminal backbone amino group and the glutamic acid side chain by means of benzotriazole 1-yl-oxy-tris-pyrrolidino-phosphonium hexafluorophosphate, hydroxybenzotriazole and N,N-diisopropylethyl amine overnight. Following cyclization, the lysine ε-amino groups were orthogonally deprotected and acylated. An individual control peptide, cyclo[Lys(Ac-Phe)-Lys(Ac-Phe)-Lys(Ac-Phe)-Glu]-Gly-OH, was synthesized along with the library. Besides serving as a control peptide for the course of the library synthesis, this peptide was also used to illustrate the enzymatic stability of the library components.

1.3.3 Chemical Transformation of Peptide Libraries

In order to enhance the physico-chemical properties of peptides and broaden their range of utility (i.e., enhanced resistance to proteolytic enzymes, high water solubility, favorable aqueous/organic partitioning characteristics, etc.), the development of soluble peptidomimetic libraries was of high interest. Advances in synthesis technologies and chemistries have led to the generation of libraries and collections of peptidomimetic and non-peptidic compounds (Gordon et al., 1994; Zuckermann, 1993).

Recently this laboratory developed an efficient method for the generation of peptidomimetic and chemical combinatorial libraries by chemically transforming existing peptide SCLs, termed the "libraries from libraries" concept (Ostresh et al., 1994a; Houghten et al., 1995). For example, the permethylation of the amide nitrogens of peptide SCLs can be carried out while they remain attached to the solid support (Ostresh et al., 1994a). This approach capitalizes on well established solid phase synthesis methods (i.e., peptide synthesis) for the straightforward preparation of a library, and enables a wide range of useful chemical diversities to be envisioned. The defining requirement of this approach is its ability to effectively transform a pool of diverse chemical moieties in a clearly understandable manner (it should be noted that while quantitative transformations are not required, reproducibility is essential). The permethylation of resin-bound peptides was carried out using a 10-fold excess of sodium hydride in dimethyl sulfoxide (DMSO) over the reactive sites for 16 hours at room tem-

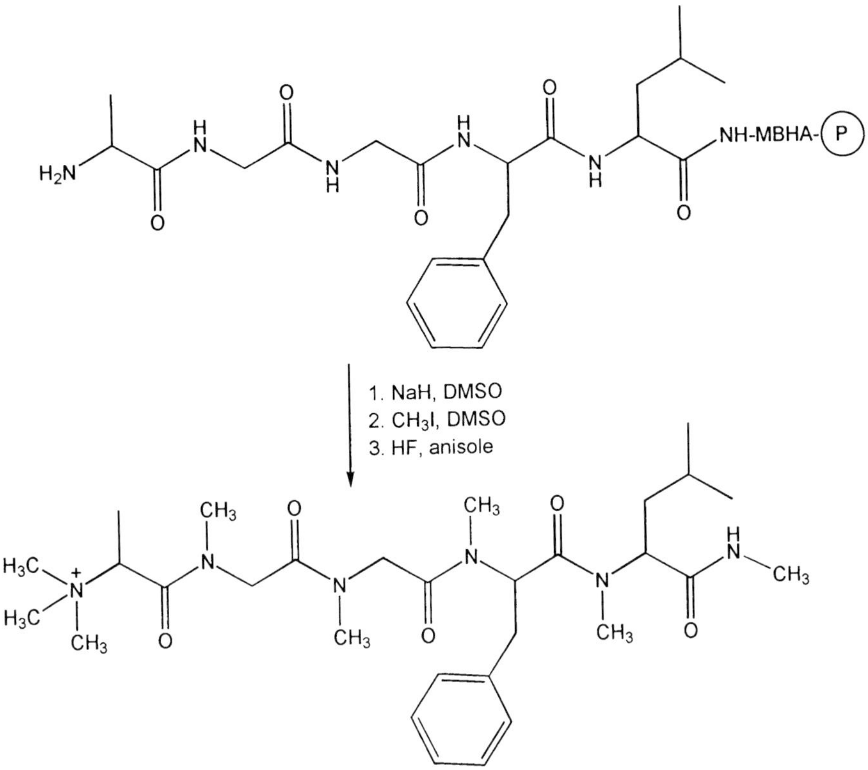

Figure 1.3: Permethylation as a method for the chemical transformation of resin-bound peptides. As an example, the permethylation of resin-bound AGGFL is shown using sodium hydride in dimethyl sulfoxide with subsequent iodomethane treatment and release of the permethylated compound from the resin using hydrogen fluoride (Ostresh et al., 1994a).

perature, followed by a 15 minute treatment of the resulting amide anions with a 30-fold excess of iodomethane over the reactive sites The synthesis scheme is shown in Figure 1.3. Under these experimental conditions, the permethylation and subsequent cleavage of 20 model peptide resins (represented by OGGFL-resin) yielded the desired products in approximately 90% purity for bifunctional amino acids. For trifunctional amino acids, side chain modifications (methylation) were reproducible. Based on these initial studies, a library resulting from permethylating the amide nitrogens of an existing hexapeptide positional scanning SCL has been successfully generated (Ostresh et al., 1994a).We have extended this method to the preparation of peralkylated libraries derived from related chemical modifications using a variety of alkylating reagents such as allyl bromide or benzyl bromide (Dörner et al., 1995). Further extension of the "libraries from libraries" concept was also achieved by chemically transforming peptide libraries by exhaustive reduction, which generates libraries of polyfunctional amines. A reproducible approach for the reduction of non-support bound peptides using 1 M borane in tetrahydrofuran has also been presented (Cuervo et al., 1995).

1.4 SCLs in Drug Discovery and Basic Research

The soluble peptide SCLs described to date range from three to 10 amino acids in length, and the number of peptides per library varies from thousands to trillions. Using either an iterative selection and enhancement process or the positional scanning format, soluble SCLs have been used successfully for the identification of: antigenic determinants (Houghten et al., 1991; Pinilla et al., 1994a; Pinilla et al., 1992; Houghten et al., 1992; Appel et al., 1992; Pinilla et al., 1993); compounds binding to opioid receptors (Pinilla et al., 1994a; Dooley and Houghten, 1993; Dooley et al., 1993); new antimicrobials (Houghten et al., 1991; Houghten et al., 1992; Blondelle and Houghten, 1994; Houghtenet al., 1993); new inhibitors of enzymes (Pinilla et al., 1994a; Eichler and Houghten, 1993) or of lytic compounds (Pinilla et al., 1994a; Blondelle and Houghten, 1994); as well as peptides having catalytic activities (Blondelle et al., 1995a). Representative examples of SCLs used in this laboratory are described below.

1.4.1 Dual Positional Hexapeptide SCL/Initial SCL

The first SCL, described above, was used for the identification of antifungal hexapeptides against *Candida albicans* (Figure 1.4). The most active individual peptide, Ac RRWWRR-NH_2, exhibits an IC_{50} of 28 µg/l (Blondelle et al., 1995a). A number of hexapeptide sequences have been identified in a similar manner which have potent activity against gram-positive bacteria such as *Staphylococcus aureus* and *Streptococcus sanguis* (Houghten et al., 1991; Houghten et al., 1992), and/or against gram-negative bacteria such as *Escherichia coli* and *Pseudomonas aeruginosa* (Houghten et al., 1993). This example not only shows that, due to their soluble character, SCLs can be used directly in cell-based assays, but also that micromolar activities can be detected from a library representing a large diversity of compounds. Although the concentration of an individual peptide in each peptide mixture of the present hexapeptide SCL is lower than micromolar range under the conditions tested (2.5 mg/ml peptide mixture screening concentration, i.e., approximately 20 nM for each individual peptide), the presence of analogs of the active sequences which show similar or lower activities within the same peptide mixture increases the actual active concentration which can be detected using standard bioassays.

1.4.2 Decapeptide PS-SCL

The positional scanning format was used to prepare a decapeptide PS-SCL made up of 10 positional libraries, each composed of 20 peptide mixtures having a single position defined and nine positions as mixtures. Each peptide mixture is made up of approximately 200 billion (2 x 10^{11}) individual sequences. Thus, each positional set of 20 peptide mixtures, as well as the entire peptide library, is composed of approximately four trillion (4 x 10^{12}) decapeptides.

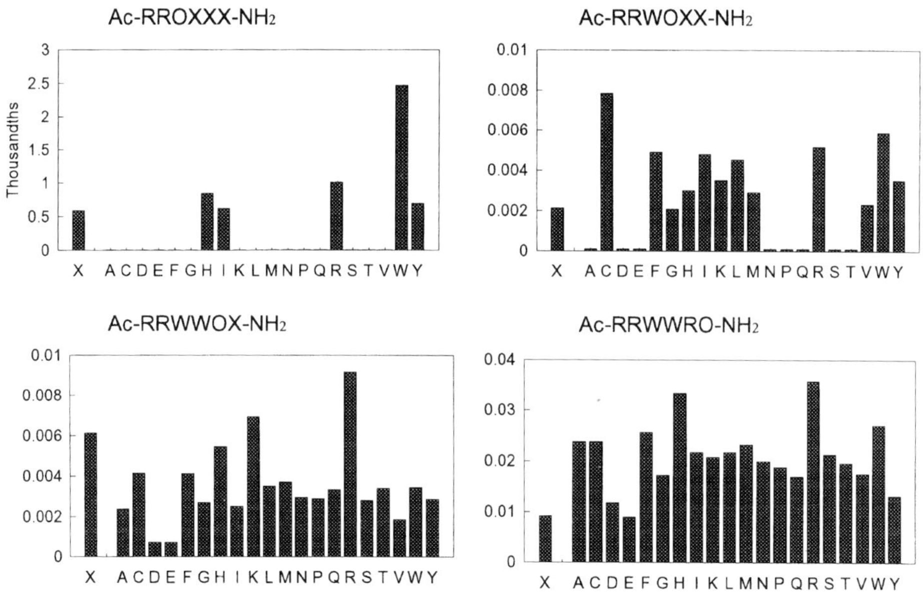

Figure 1.4: Identification of antifungal hexapeptides (Blondelle et al., 1995a). The activity was determined against 10^5 CFU/ml *Candida albicans* following a 48 h incubation at 30° C as described elsewhere (Blondelle et al., 1994). The IC_{50} values (in µg/ml – peptide mixture concentration necessary to inhibit 50% cell growth) were determined from two-fold serial dilutions of the peptide mixtures, and calculated using a sigmoidal curve fitting method. Each graph represents each step carried out during the iterative process from the peptide mixture Ac-RRXXXX-NH_2. Each bar on the y-axis represents the reciprocal of the IC_{50} values, and is labeled on the x-axis by the amino acid used to define the "O" position. X represents the peptide mixture from the previous iterative step.

Table 1.3: Inhibition of mAb 17/9 by the 21 most effective decapeptides derived from the PS-SCL

#	Sequence	IC_{50} (nM)
1	Ac-DDDDDVPDYA-NH_2	0.6
2	Ac-DDDDDVADYA-NH_2	0.8
3	Ac-DDDDDVEDYA-NH_2	1.1
4	Ac-DDDVDVPDYA-NH_2	1.1
5	Ac-DDDVDVADYA-NH_2	2.2
6	Ac-DDDDDVDDYA-NH_2	2.8
7	Ac-DDDVDVEDYA-NH_2	3.3
8	Ac-DDDVDDPDYA-NH_2	4.4
9	Ac-DDDVDDYAAA-NH_2	6.3
10	Ac-DDDVDVYDYA-NH_2	7.6
11	Ac-DDDDDVYDYA-NH_2	7.8
12	Ac-DDDDDDPDYA-NH_2	8.0
13	Ac-DDDVDDEDYA-NH_2	8.7
14	Ac-DDDDDDEDYA-NH_2	8.7
15	Ac-DDDVDVDDYA-NH_2	9.5
16	Ac-DDDDDDADYA-NH_2	38
17	Ac-DDDVDDYADA-NH_2	45
18	Ac-DDDVDDADYA-NH_2	62
19	Ac-DDDDDVDYAA-NH_2	77
20	Ac-DDDVDVDYAA-NH_2	83
21	Ac-DDDVDDYAYA-NH_2	85
control	Ac-YPYDVPDYASLRS-NH_2	6

In the first example presented (Pinilla et al., 1994b), this library was screened against a monoclonal antibody raised against a 13-mer peptide (mAb 17/9), the specificity of which had been previously characterized by competitive ELISA and x-ray structural methods using substitution analogs. Individual decapeptides were prepared from combinations of the most effective amino acid residue from each of the 10 positions. Twenty-one decapeptides (Table 1.3; Pinilla et al., 1994b), all containing the specific motif -DYA-, had IC_{50} values (the concentration of peptide mixture necessary to inhibit 50% antibody binding) lower than 100 nM, 17 of which contain this sequence motif in positions 8–10. Four of these peptides, all of which contain multiple aspartic acids in the N-terminal region, were found to be nearly five times more effective than the 13-residue control peptide Ac-YPYDVPDYASLRS-NH_2 (IC_{50} = 6nM).

More recently, this decapeptide library was screened against three different monoclonal antibodies (Appel et al., 1995). Individual decapeptides corresponding to sequences derived from the most active peptide mixtures at each position were synthesized and assayed. In one case, the most active individual decapeptide having a multiple aspartic acid motif added to the specific antigenic determinant improved recognition 100-fold. In two other examples, decapeptides having high affinities comparable to the original immunogens were identified. Despite the immense number of decapeptides, specific sequences having nanomolar affinities have been identified for four different antibodies (Pinilla et al., 1994b; Appel et al., 1995).

1.4.3 SCLs Composed of L-, D-, and/or Unnatural Amino Acids

While L-amino acid peptides are expected to be susceptible to proteolytic breakdown and, in turn, may be limited to topical therapeutic use, the insertion of D- and/or unnatural amino acids into a sequence is anticipated to increase the duration of activity and applicability of such compounds.

An acetylated hexapeptide library composed entirely of D-amino acids, represented as Ac-$o_1o_2$xxxx-NH_2, has been used for the identification of a novel μ-specific opioid peptide (Dooley et al., 1994). This library was screened for its ability to inhibit the α binding of [^{3}H][D-Ala^2,MePhe^4,Gly^5-ol] enkephalin (DAMGO) to rat brain homogenates. The most active peptide found upon completion of the iterative process (Figure 1.5) was Ac-rfwink-NH_2 (IC_{50} = 18nM). This peptide was found to be an agonist with high selectivity for the μ receptor (Dooley and Houghten, 1994). Furthermore, this peptide produces potent analgesia in mice by intracerebroventricular or intraperitoneal administration (Dooley et al., 1994). It should be noted that both antagonists and agonists can be identified through the screening of an SCL. For instance, novel all L-amino acid peptides having fully antagonist properties at the μ receptor were identified from an acetylated hexapeptide SCL [termed acetalins (Dooley et al., 1993)], while the

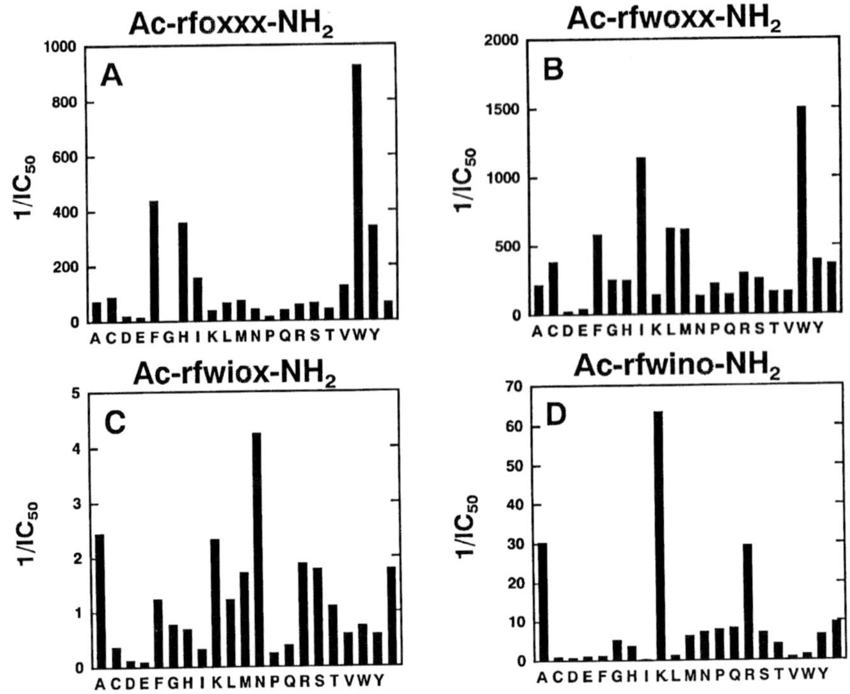

Figure 1.5: Identification of an all D-amino acid opioid peptide. The library was screened for ability to inhibit the binding of 7nM ^{3}H-DAMGO to crude rat brain homogenates (Dooley et al., 1994). Each graph represents a screening step carried out during the iterative process from the peptide mixture Ac-rfxxxx-NH_2. The IC_{50} values represent the concentration necessary to inhibit 50% binding of ^{3}H-DAMGO; the x- and y-axis are defined as described in Figure 4. Lower case letters represent D-amino acids.

screening of the corresponding D counterpart SCL led to the identification of an agonist.

The screening of an N-acetylated hexapeptide D-amino acid PS-SCL for trypsin inhibition (Pinilla et al., 1994a) also led to active peptides. In this case, since specific answers could not be determined from all the single positional libraries, an iterative process was initiated from the most specific positions. Ac-ryrpwp-NH_2 ($IC_{50} = 62 \mu M$) was identified as the D-amino acid hexapeptide with the strongest trypsin inhibitory activity. This peptide was shown to be stable towards tryptic hydrolysis. This example illustrates the fact that the positional SCLs making up a PS-SCL can be considered independent of each other, and can each be used as a starting point for an iterative synthesis and screening process as described for dual positional SCLs (Houghten et al., 1991).

In order to increase the molecular diversity of the libraries, a tetrapeptide library composed of L-, D- and unnatural amino acids (UZZZ-NH_2) was prepared and screened in an opioid receptor assay. The first position of this library was defined with one of 58 different amino acids (U: 20 L-, 19 D- and 19 unnatural amino acids; listed in Table 1.4) and the three remaining positions close to equimolar mixtures of 56 of the same amino acids (Z: L- and D-cysteine were omitted). The YZZZ-NH_2 mixture showed the most effective inhibition (Dooley et al., 1995). Upon completing the iterations, the most active mixtures found were YmFA-NH_2, YmF(-Aib)-NH_2 (α-Aib: α-aminoisobutyric acid) and YmF(p-nitro)F-NH_2 (IC_{50} = 2 nM, 2 nM and 4 nM, respectively). The peptide YmFH (IC_{50} = 38 nM), also identified from this library, represents the first four residues of Dermenkephaline (Charpentier et al., 1991). Other peptides containing the YmF-motif have also been described for their opioid activity (Lazarus et al., 1991).

Table 1.4: Unnatural amino acids used in the tetrapeptide SCL

Protected amino acid derivative used during the synthesis
N-Boc-β-alanine
N-Boc-L-α-aminobutyric acid
N-Boc-γ-aminobutyric acid
N-Boc-α-aminoisobutyric acid
N-Boc-ε-aminocaproic acid
N-Boc-7-aminoheptanoic acid
N-Boc-L-aspartic acid α-benzyl ester
N-Boc-L-glutamic acid-α-benzyl ester
N-Boc-S-acetamidomethyl-L-cysteine
N-ε-Boc-N-α-Cbz-L-lysine
N-ε-Boc-N-α-Fmoc-L-lysine
N-Boc-L-methionine sulfone
N-Boc-L-norleucine
N-Boc-L-norvaline
N-α-Boc-N-δ-Cbz-L-ornithine
N-δ-Boc-N-α-Cbz-L-ornithine
N-Boc-p-nitro-L-phenylalanine
N-Boc-O-benzyl-L-hydroxyproline
N-Boc-L-thioproline

Cbz: Benzyloxycarbonyl

Since antimicrobial peptides composed of L-amino acids are susceptible to proteolytic breakdown and therefore limited to topical or intravenous use, the two SCLs (UZZZ-NH_2 and Ac-UZZZ-NH_2) were also assayed for inhibition of *S. aureus* (Blondelle et al., 1994) and *E. coli* (Blondelle and Houghten, 1994). (αFmoc-εLys)ZZZ-NH_2 showed the highest activity in both assays ($IC_{50} = 44$ and 179 μg/ml, respectively). Following an iterative pro-cess, individual tetrapeptides having antimicrobial activities were identified (Table 1.5; Blondelle et al., 1994). These tetrapeptides also exhibited a broad spectrum of activities as determined against five different microorganisms (*S. aureus*, methicillin resistant *S. aureus* (MRSA), *S. sanguis*, *E. coli*, and *C. albicans*) (Blondelle et al., 1994).

Table 1.5: Antimicrobial activities of tetrapeptides derived from (αFmoc-εLys)ZZZ-NH_2

Sequence	MIC (μg/ml)	
	S. aureus	*E. coli*
(αFmoc-εLys)WfR-NH_2	4– 8	31– 62
(αFmoc-εLys)Wfl-NH_2	3– 4	62–125
(αFmoc-εLys)Wf(Thiopro)-R-NH_2	4– 8	31– 62
(αFmoc-εLys)WKW-NH_2	5– 8	16– 31
(αFmoc-εLys)WKw-NH_2	8–16	31– 62
(αFmoc-εLys)WK(NO_2F)-NH_2	8–16	31– 62
(αFmoc-εLys)WYr-NH_2	5– 8	31– 62
(αFmoc-εLys)WY(αAba)-NH_2	5– 8	62–125
(αFmoc-εLys)cir-NH_2	4– 8	31– 62
(αFmoc-εLys)ciK-NH_2	8–16	62–125
(αFmoc-εLys)ci(δOrn)-NH_2	8–16	31– 62

Thiopro: L-thioproline; αAba: L-α-aminobutyric acid; δOrn: L-δ-ornithine; NO_2F: p-nitro-L-phenylalanine. MIC: Minimum inhibitory concentration.

1.4.4 Cyclic Peptide Template Library

The reduced flexibility of cyclic peptides is thought to increase their potential for high affinity binding to various acceptor molecules. Furthermore, template molecules are expected to represent "surface mimetics" of discontinuous binding sites of proteins. Since enzyme inhibitors have a wide range of research and therapeutic uses, the positional scanning cyclic template library composed of three positional libraries was screened using chymotrypsin as a model target (Eichler et al. 1994b; Eichler et al., 1995).

The cyclic peptide mixtures were separately tested for their ability to inhibit the chymotryptic hydrolysis of the chromogenic substrate N-succinyl-L-phenylalanine-p-nitroanilide. The most effective functionalities at each of the three lysine side chains were found to be piperonylic acid and 2-thiophenecarboxylic acid. The inhibitory activity of cyclic compounds with these functionalities ranged from IC_{50} = 51 μM to IC_{50} = 94 μM. The most active chymotrypsin inhibitor is shown in Figure 1.6. It should be noted that this cyclic compound has no structural resemblance to any known natural or synthetic chymotrypsin inhibitors, indicating the potential of SCLs for the identification of novel enzyme inhibitors.

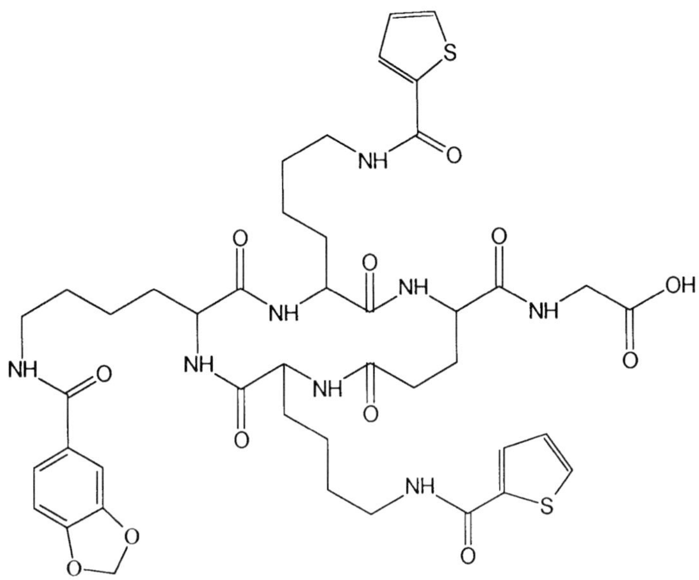

Figure 1.6: Chymotrypsin inhibitor found upon screening a cyclic peptide template combinatorial library. The inhibitory activity found for this compound was IC_{50} = 51 μM.

1.4.5 Transformed Peptide Libraries

The general poor oral availability and rapid breakdown of linear L-amino acid peptides make them unlikely drug candidates as compared to organic molecules. Since peptide SCLs have proven to be useful tools for rapid drug discovery, this field has currently moved its emphasis toward peptidomimetic and chemical libraries. The first library composed of peptidomimetics that has been screened consisted of permethylated compounds in a positional scanning format (Ostresh et al., 1994a). This permethylated PS-SCL was screened for inhibition of antimicrobial activity against *S. aureus* growth (Ostresh et al., 1994a; Houghten et al., 1995). The permethylated mixtures having a hydrophobic residue or the permethylated form of histidine at the defined position showed the highest activity at each position. A set of 144 individual peptides representing all combinations of the most active residues was then generated and permethylated to be assayed for inhibition of S. aureus. The most active permethylated compounds from this set (Table 1.6) were pm[LFIFFF NH_2], pm[FFIFFF-NH_2], pm[FFFFFF-NH_2] and pm[LFFFFF NH_2] (IC_{50} = 6, 6, 7 and 10 µg/ml, respectively). These permethylated compounds showed similar activity against MRSA and *S. sanguis,* and no activity against the gram negative bacteria *E. coli* and yeast *C. albicans.*

Since phenylalanine was active at every position of the permethylated PS-SCL, a series of phenylalanine-containing peptides, which ranged in length from a single phenylalanine amide to its octapeptide form, was then synthesized and permethylated. This series was prepared in order to determine if the repetitive appearances of phenylalanine found at all six positions in the permethylated PS-SCL were due to an individual hexamer sequence, or to a frame-shifting fit by a shorter sequence (i.e., permethylated di-, tri-, tetra-phenylalanine). The highest activity found (MIC = 2.5 µg/ml; minimum inhibitory concentration: lowest concentration at which no growth is detected after 21 hour incubation) was associated with the permethylated heptamer sequence.

Table 1.6: Antimicrobial and hemolytic activities of permethylated peptides derived from a permethylated SCL

Parent	S. aureus		MRSA		S. sanguis		%hemolysis
Sequence	IC_{50}(µg/ml)	MIC(µg/ml)	IC_{50}(µg/ml)	MIC(µg/ml)	IC_{50}(µg/ml)	MIC(µg/ml)	at 350 µg/ml
LFIFFF-NH_2	6	11–15	7	8–10	17	3–5	12
FFIFFFNH_2	6	11–15	7	8–10	14	20–40	0
FFFFFF-NH_2	7	11–15	7	8–10	9	15–20	16
LFFFFF-NH_2	10	21–31	8	9–10	14	20–40	10
FFFFHF-NH2	11	15–21	18	21–42	19	30–40	3
LFIFFH-NH_2	12	21–31	20	25–42	19	30–40	1
LFFFHF-NH_2	12	21–31	13	15–21	19	30–40	4
LFIFHF-NH_2	12	21–31	14	15–21	13	20–40	7
LFFFFH-NH_2	13	21–42	16	21–42	22	30–40	3
FFIFFH-NH_2	14	31–42	19	21–42	18	30–40	0

1.4.6 Design of Synthetic "Enzymes"

The use of SCLs to generate immensely diverse planer or topographical landscapes is now possible. Thus, one can readily envision the design of fully synthetic "bioreceptors" which are custom designed to bind to specific ligands of virtually any type, as well as the custom design of "enzymes" (i.e., complexes having catalytic properties) able to cleave or combine a wide variety of building blocks. Artificial enzymes and/or receptors can be envisioned to fulfill a wide range of basic research, therapeutic and industrial needs. Such libraries must fulfill the minimum structural requirement necessary to mimic the natural "enzymes" or "bioreceptors".

A fundamental characteristic of biological functions such as biocatalysis and molecular recognition is that, upon binding of a macromolecule to a bioreceptor, the relevant function takes place in a conformationally restricted environment. We have successfully designed conformationally defined libraries consisting of combinatorial mixtures of building blocks inserted in a structurally defined peptide sequence (Blondelle et al., 1995a; Blondelle et al., 1995b). This new structural approach has been used to identify peptides that catalyze hydrolysis and decarboxylation reactions such as the decarboxylation of oxaloacetate. Using a conformationally defined SCL derived from an amphipathic -helical 18-mer peptide (Blondelle and Houghten, 1992), we have identified a set of individual peptides with substantially higher catalytic activity than the recently reported catalytic peptide oxaldie-1 (Johnsson et al., 1993). Thus, YKLLKELLAKLKWLLRKL-NH_2 was found to catalyze the decarboxylation of oxaloacetate with a catalytic constant $k_{cat} = 1.5 \times 10^{-2} s^{-1}$. This reaction proceeds at a rate 10^3- to 10^4-fold faster than the simple amine catalysts, and was found to correlate well with the ability of this peptide to fold into a defined a-helical conformation.

The recent development of conformationally defined libraries opens a new phase in the combinatorial library field for applications in molecular biology and chemistry. Although still in the beginning stages, the promising results reported indicate that artificial receptors, enzymes, templates, and self assembling proteins will soon be available for a broad cross section of applications.

1.5 Conclusion

The generation of soluble SCLs comprised of tens to hundreds of millions of peptides, peptidomimetics and organic compounds can be now carried out with a high degree of confidence and exactitude [DeWitt et al., 1993; Blake and Litzidavis, 1992; Eichler et al., 1994b; Zuckermann, 1993]. Such libraries have proven to be widely useful for the identification of a variety of biologically active compounds. Since the majority of pharmacologically relevant assays involve membrane-bound receptors (e.g., bacterial and viral cell membranes, etc.), solubility is a key factor in simplifying drug screening. In order to fulfill all the requirements of new drug candidates (high stability to enzymatic degradation, oral availability, etc.) future studies can be expected to be directed toward the preparation of large chemical combinatorial libraries. The libraries currently available, as well as the development of organic compound libraries, will play an increasingly important role in the search for novel pharmacophores and in the development of new methods useful in all areas of basic research.

Acknowledgements

We thank Eileen Silva for her assistance in preparing this manuscript. This work was funded in part by Houghten Pharmaceuticals, Inc., San Diego, California.

References

Appel, J.R., Pinilla, C., and Houghten, R.A. (1992). Identification of related peptides recognized by a monoclonal antibody using a synthetic peptide combinatorial library. Immunomethods *1,* 17–23.

Appel, J.R., Buencamino, J., Houghten, R.A., and Pinilla, C. (1995). Study of peptide-antibody interactions using a decapeptide positional scanning library. In: Peptides 94: Proceedings of the 23rd European Peptide Symposium, H.L.S. Maia, ed. (Leiden: Escom), pp. 815–816.

Blake, J., and Litzidavis, L. (1992). Evaluation of peptide libraries – an iterative strategy to analyze the reactivity the reactivity of peptide mixtures with antibodies. Bioconjugate Chem. *3,* 510–513.

Blondelle, S.E., and Houghten, R.A. (1992). Design of model peptides having potent antimicrobial activities. Biochemistry *31,* 12688–12694.

Blondelle, S.E., and Houghten, R.A. (1994). Membrane protecting sequences and new antimicrobial peptides identified through the screening of synthetic peptide combinatorial libraries. In: Techniques in Protein Chemistry V, J.W. Crabb, ed. (Orlando: Academic Press), pp. 509–516.

Blondelle, S.E., Takahashi, E., Weber, P.A., and Houghten, R.A. (1994). Identification of antimicrobial peptides using combinatorial libraries made up of unnatural amino acids. Antimicrob. Agents Chemother. *38,* 2280–2286.

Blondelle, S.E., Pérez-Payá, E., Dooley, C.T., Pinilla, C., and Houghten, R.A. (1995a). Chemical combinatorial libraries, peptidomimetics and peptide diversity. Trends Anal. Chem. *14,* 83–92.

Blondelle, S.E., Takahashi, E., Houghten, R.A., and Pérez-Payá, E. (1995b) Design of conformationally defined combinatorial libraries. In: Peptides 94: Proceedings of the 23rd European Peptide Symposium, H.L.S. Maia, ed. (Leiden: Escom), pp. 85–86.

Bunin, B.A., Plunkett, M.J., and Ellman, J.A. (1994). The combinatorial synthesis and chemical and biological evaluation of a 1,4-benzodiazepine library. Proc. Natl. Acad. Sci. USA *91,* 4708–4712.

Charpentier, S., Sagan, S., Delfour, A., and Nicolas, P. (1991). Dermenkephalin and deltorphin I reveal similarities within ligand-binding domains of μ- and δ-opioid receptors and an additional address subsite on the d-receptor. Biochem. Biophys. Res. Commun. *179,* 1161–1168.

Cho, Y.C., Moran, E.J., Cherry, S.R., Stephans, J.C., Fodor, S.P.A., Adams, C.L., Sundaram, A., Jacobs, J.W., and Schultz, P.G. (1993). An unnatural biopolymer. Science *261,* 1303–1305.

Cuervo, J. H., Weitl, F., Ostresh, J. M., Hamashin, V. T., Hannah, A. L., and Houghten, R. A. (1995). Polyalkylamine chemical combinatorial libraries. In: Peptides *94:* Proceedings of the 23rd European Peptide Symposium, H. L. S. Maia, ed. (Leiden: Escom), pp. 465–466.
Cwirla, S. E., Peters, E. A., Barrett, R. W., and Dower, W. J. (1990). Peptides on phage: A vast library of peptides for identifying ligands. Proc. Natl. Acad. Sci. USA *87,* 6378–6382.
Darlak, K., Romanovskis, P., and Spatola, A. F. (1994). Cyclic Peptide Libraries. In: Peptides, Proceedings of the 13th American Peptide Symposium, R.S. Hodgesand J.A. Smith, eds. (Leiden: ESCOM), pp. 981–983.
Devlin, J. J., Panganiban, L. C., and Devlin, P. E. (1990). Random peptide libraries: A source of specific protein binding molecules. Science *249,* 404–406.
DeWitt, S. H., Kiely, J. S., Stankovic, C. J., Schroeder, M. C., Cody, D. M. R., and Pavia, M. R. (1993). "Diversomers": An approach to nonpeptide, nonoligomeric chemical diversity. Proc. Natl. Acad. Sci. USA *90,* 6909–6913.
DiMaio, J., and Schiller, P.W. (1980). A cyclic enkephalin analog with high in vitro opiate activity. Proc. Natl. Acad. Sci. USA *77,* 7162–7166.
Dooley, C.T., and Houghten, R. A. (1993). The use of positional scanning synthetic peptide combinatorial libraries for the rapid determination of opioid receptor ligands. Life Sciences *52,* 1509–1517.
Dooley, C. T., Chung, N. N., Schiller, P. W., and Houghten, R. A. (1993). Acetalins: New opioid receptor antagonists determined through the use of Synthetic Peptide Combinatorial Libraries. Proc. Natl. Acad. Sci. USA *90,* 10811–10815.
Dooley, C. T., and Houghten, R. A. (1994). New, potent N-acetylated all D-amino acid opioid peptides. In: Peptides: Chemistry, Structure and Biology. Proceedings of the 13th American Peptide Symposium, R. S. Hodgesand J. A. Smith, eds. (Leiden: Escom), pp. 984–985.
Dooley, C. T., Hope, S. K., and Houghten, R. A. (1995). Identification of tetrameric opioid peptides from a combinatorial library composed of L-, D- and non-proteinogenic amino acids. In: Peptides 94: Proceedings of the 23rd European Peptide Symposium, H. L. S. Maia, ed. (Leiden: Escom), pp. 805–806.
Dooley C. T, Chung N. N, Wilkes B. C, Schiller, P. W., Bidlack, J. M., Pasternak, G. W., and Houghten, R. A. (1994). An all D-amino acid opioid peptide with central analgesic activity from a combinatorial library. Science, *266,* 2019–2022.
Dörner, B., Ostresh, J. M., Husar, G. M., and Houghten, R. A. Extending the range of molecular diversity through amide alkylation of peptide libraries. In: Peptides 94: Proceedings of the 23rd European Peptide Symposium, H. L. S. Maia, ed. (Leiden: Escom), in press.
Eichler, J., and Houghten, R. A. (1993). Identification of substrate-analog trypsin inhibitors through the screening of synthetic peptide combinatorial libraries. Biochemistry *32,* 11035–11041.
Eichler, J., Lucka, A. W., and Houghten, R. A. (1994b). Cyclic peptide template combinatorial libraries: Synthesis and identification of chymotrypsin inhibitors. Pept. Res. *7,* 300–307.
Eichler, J., Pinilla, C., Chendra, S., Appel, J. R., and Houghten, R. A. (1994a). Synthesis of peptide libraries on cotton carriers: methods and applications. In: Innovation and Perspectives in Solid Phase Synthesis – Peptides, Polypeptides and Oligonucleotides, R. Epton, ed. (Birmingham: Mayflower Worldwide Limited), pp. 227–232.
Eichler, J., Lucka, A. W., and Houghten, R. A. (1995). Cyclic peptide template synthetic combinatorial libraries. In: Peptides 94: Proceedings of the 23rd European Peptide Symposium, H. L. S. Maia, ed. (Leiden: Escom), pp. 461–462.
Fodor, S. P. A., Read, J. L., Pirrung, M. C., Stryer, L., Lu, A. T., and Solas, D. (1991). Light-directed, spatially addressable parallel chemical synthesis. Science *251,* 767–773.
Furka, A., Sebestyen, F., Asgedom, M., and Dibo, G. (1991). General method for rapid synthesis of multicomponent peptide mixtures. Int. J. Pept. Protein Res. *37,* 487–493.

Gallop, M.A., Barrett, R.W., Dower, W.J., Fodor, S.P.A., and Gordon, E.M. (1994). Applications of combinatorial technologies to drug discovery. 1. Background and peptide combinatorial libraries. J. Med. Chem. *37,* 1233–1251.

Geysen, H.M., Rodda, S.J., and Mason, T.J. (1986). A priori delineation of a peptide which mimics a discontinuous antigenic determinant. Mol. Immunol. *23,* 709–715.

Geysen, H.M., and Mason, T.J. (1993). Screening chemically synthesized peptide libraries for biologically-relevant molecules. BioMed. Chem. Lett. *3,* 397–404.

Gordon, E.M., Barrett, R.W., Dower, W.J., Fodor, S.P.A., and Gallop, M.A. (1994). Applications of combinatorial technologies to drug discovery. 2. Combinatorial organic synthesis, library screening strategies, and future directions. J. Med. Chem. *37,* 1385–1401.

Hortin, G.L., Staatz, W.D., and Santoro, S.A. (1992). Preparation of soluble peptide libraries: Application to studies of platelet adhesion sequences. Biochem. Int. *26,* 731–738.

Houghten, R.A. (1985). General method for the rapid solid-phase synthesis of large numbers of peptides: specificity of antigen antibody interaction at the level of individual amino acids. Proc. Natl. Acad. Sci. USA *82,* 5131–5135.

Houghten, R.A., Bray, M.K., De Graw, S.T., and Kirby, C.J. (1986). Simplified procedure for carrying out simultaneous multiple hydrogen fluoride cleavages of protected peptide resins. Int. J. Pept. Protein Res. *27,* 673–678.

Houghten, R.A., Pinilla, C., Blondelle, S.E., Appel, J.R., Dooley, C.T., and Cuervo, J.H. (1991). Generation and use of synthetic peptide combinatorial libraries for basic research and drug discovery. Nature *354,* 84–86.

Houghten, R.A., Appel, J.R., Blondelle, S.E., Cuervo, J.H., Dooley, C.T., and Pinilla, C. (1992). The use of synthetic peptide combinatorial libraries for the identification of bioactive peptides. Biotechniques *13,* 412–421.

Houghten, R.A., Dinh, K.T., Burcin, D.E., and Blondelle, S.E. (1993). The systematic development of peptides having potent antimicrobial activity against *E. coli* through the use of synthetic peptide combinatorial libraries. In: Techniques in Protein Chemistry IV, R.H. Angeletti, ed. (Orlando: Academic Press), pp. 249–256.

Houghten, R.A. (1994). Finding the needle in the haystack. Curr. Biol. *4,* 564–567.

Houghten, R.A., Ostresh, J.M., Husar, G.M., Dörner, B., and Blondelle, S.E. (1995). Libraries from libraries: The generation of peptidomimetic combinatorial diversities. In: Peptides 94: Proceedings of the 23rd European Peptide Symposium, H.L.S. Maia, ed. (Leiden: Escom), pp. 459–460.

Johnsson, K., Allemann, R.K., Widmer, H.,and Benner, S.A. (1993). Synthesis, structure and activity of artificial, rationally designed catalytic polypeptides. Nature *365,* 530–532.

Kerr, J.M., Banville, S.C., and Zuckermann, R.N. (1993). Encoded combinatorial peptide libraries containing non-natural amino acids. J. Am. Chem. Soc. *115,* 2529–2531.

Lam, K.S., Salmon, S.E., Hersh, E.M., Hruby, V.J., Kazmierski, W.M., and Knapp, R.J. (1991). A new type of synthetic peptide library for identifying ligand-binding activity. Nature *354,* 82–84.

Lam, K.S., and Lebl, M. (1992). Streptavidin and avidin recognize peptide ligands with different motifs. Immunomethods *1,* 11–15.

Lazarus, L.H., Salvadori, S., Santagada, V., Tomatis, R., and Wilson, W.E. (1991). Function of negative charge in the "Address Domain" of deltorphins. J. Med. Chem. *34,* 1350–1355.

Needels, M.C., Jones, D.G., Tate, E.H., Heinkel, G.L., Kochersperger, L.M., Dower, W.J., Barrett, R.W., and Gallop, M.A. (1993). Generation and screening of an oligonucleotide-encoded synthetic peptide library. Proc. Natl. Acad. Sci. USA *90,* 10700–10704.

Nikolaiev, V., Stierandová, A., Krchnák, V., Seligmann, B., Lam, K.S., Salmon, S.E., and Lebl, M. (1993). Peptide-encoding for structure determination of nonsequenceable polymers within libraries synthesized and tested on solid-phase supports. Pept. Res. *6,* 161–170.

Ohlmeyer, M.H.J., Swanson, R.N., Dillard, L.W., Reader, J.C., Asouline, G., Kobayashi, R., Wigler, M., and Still, W.C. (1993). Complex synthetic chemical libraries indexed with molecular tags. Proc. Natl. Acad. Sci. USA *90*, 10922–10926.

Ostresh, J.M., Husar, G.M., Blondelle, S.E., Dörner, B., Weber, P.A., and Houghten, R.A. (1994a). "Libraries from libraries": Chemical transformation of combinatorial libraries to extend the range and repertoire of chemical diversity. Proc. Natl. Acad. Sci. USA *91*, 11138–11142.

Ostresh, J.M., Winkle, J.H., Hamashin, V.T., and Houghten, R.A. (1994b). Peptide libraries: Determination of relative reaction rates of protected amino acids in competitive couplings. Biopolymers *34*, 1681–1689.

Owens, R.A., Gesellchen, P.D., Houchins, B.J., and DiMarchi, R.D. (1991). The rapid identification of HIV protease inhibitors through the synthesis and screening of defined peptide mixtures. Biochem. Biophys. Res. Comm. *181*, 402–408.

Pinilla, C., Appel, J.R., Blanc, P., and Houghten, R.A. (1992). Rapid identification of high affinity peptide ligands using positional scanning synthetic peptide combinatorial libraries. Biotechniques *13*, 901–905.

Pinilla, C., Appel, J.R., and Houghten, R.A. (1993). Synthetic peptide combinatorial libraries (SPCLs): Identification of the antigenic determinant of b-endorphin recognized by monoclonal antibody 3E7. Gene *128*, 71–76.

Pinilla, C., Appel, J.R., Blondelle, S.E., Dooley, C.T., Eichler, J., Ostresh, J.M., and Houghten, R.A. (1994a). Versatility of positional scanning synthetic combinatorial libraries for the identification of individual compounds. Drug Dev. Res. *33*, 133–145.

Pinilla, C., Appel, J.R., and Houghten, R.A. (1994b). Investigation of antigen-antibody interactions using a soluble nonsupport-bound synthetic decapeptide library composed of four trillion sequences. Biochem. J. *301*, 847–853.

Pinilla, C., Appel, J., Blondelle, S.E., Dooley, C.T., Dörner, B., Ostresh, J.M., and Houghten, R.A.(1995). A review of the utility of peptide combinatorial libraries. Biopolymers (Peptide Science) *37*, 221–240.

Scott, J.K., and Smith, G.P. (1990). Searching for peptide ligands with an epitope library. Science *249*, 386–390.

Scott, J.K., and Craig, L. (1994). Random peptide libraries. Curr. Opin. Biotechnol. 5, 40–48. Sham, H.L.Bolis, G., Stein, H.H., Fesik, S.W., Marcotte, P.A., Plattner, J.J., Rempel, C.A., and Greer, J. (1988). Renin inhibitors. Design and synthesis of a new class of conformationally restricted analogues of angiotensinogen. J. Med. Chem. *31*, 284–295.

Simon, R.J., Kania, R.S., Zuckermann, R.N., Huebner, V.D., Jewell, D.A., Banville, S., Ng, S., Wang, L., Rosenberg, S., Marlowe, C.K., Spellmeyer, D.C., Tan, R., Frankel, A.D., Santi, D.V., Cohen, F.E., and Bartlett, P.A. (1992). Peptoids: A modular approach to drug discovery. Proc. Natl. Acad. Sci. USA *89*, 9367–9371.

Smith, G.P., Schultz, D.A., and Ladbury, J.E. (1993). A ribonuclease S-peptide antagonist discovered with a bacteriophage display library. Gene *128*, 37–42.

Tam, J.P., Heath, W.F., and Merrifield, R.B. (1983). SN2 deprotection of synthetic peptides with a low concentration of HF in dimethyl sulfide: evidence and application in peptide synthesis. J. Am. Chem. Soc. *105*, 6442–6455.

Tuchscherer, G., Dörner, B., Sila, U., Kamber, B., and Mutter, M. (1993). The TASP concept: Mimetics of peptide ligands, protein surfaces and folding units. Tetrahedron *49*, 3559–3575.Zuckermann, R.N. (1993). The chemical synthesis of peptidomimetic libraries. Curr. Opin. Struct. Biol. *3*, 580–584.

Zuckermann, R.N., Martin, E.J., Spellmayer, D.C., Stauber, G.B., Shoemaker, K.R., Kerr, J.M., Figliozzi, G.M., Goff, D.A., Siani, M.A., Simon, R.J., Banville, S.C., Brown, E.G., Wang, L., Richter, L.S., and Moos, W.H. (1994). Discovery of nanomolar ligands for 7-transmembrane G-protein-coupled receptors from a diverse N-(substituted) glycine peptoid library. J. Med. Chem. *37*, 2678–2685.

2 Combinatorial Libraries of Synthetic Structures: Synthesis, Screening, and Structure Determination

Viktor Krchňák, Nikolai F. Sepetov, Petr Kocis, Marcel Patek,
Kit S. Lam, Michal Lebl

2.1 Introduction

One of the basic questions for which modern science continues to seek an answer is the nature of intermolecular communication. Many biological effects are triggered by the interaction of a ligand with its signaling counterpart, and many diseases are caused by the malfunction of this signaling mechanism. Once the role of a particular signaling mechanism is recognized, one may deliberately intercede in this process in an attempt to obviate pathological activity. If there is no *a priori* knowledge of structure activity relationships between ligand and receptor, or one would like to discover a novel structural motif, the only practical approach is to test a multitude of compounds for the desired activity. The success of this random screening approach depends on both the number and the spread of diversity among compounds available for testing.

Of the approaches to the generation of molecular diversity, multiple synthesis – in which a multitude of compounds are separately though simultaneously synthesized – was the first to accelerate dramatically the speed of compound synthesis (Geysen et al., 1984; Houghten et al., 1985; Frank et al., 1988; Krchňák and Vagner, 1990; Fodor et al., 1991; DeWitt et al., 1993; Kramer et al., 1993; Meldal et al., 1993; Bunin et al., 1994; Cass et al., 1994). However, it was the advent of combinatorial techniques (Geysen et al., 1986) that have revolutionized high throughput screening and, as such, can be considered one of the most exciting recent developments in medicinal chemistry. The first breakthrough in generating vast numbers of compounds in a format suitable for high throughput assays, in this case peptides, was the filamentous phage technique (Parmley and Smith, 1989; Cwirla et al., 1990; Devlin et al., 1990; Scott and Smith, 1990; Felici et al., 1991; Cull et al., 1992; O'Neil et al., 1994). In contrast to multiple synthesis techniques, in which the structure of the compound to be synthesized is known throughout the process, the phage technique represents a true "library," in which the structure of a tested compound is not determined until its biological relevance has been established. Soon thereafter, complex synthetic combinatorial libraries emerged, including Lam's one-bead one-structure concept (Lam et al., 1991) (based on Furka's ingenious split/mix method (Furka et al., 1988a; Furka et al., 1988b; Furka et al., 1991)) and Houghten's iterative combinatorial library strategy (Houghten et al., 1991) (analogous to Geysen's mimotope strategy (Geysen et al., 1986)). The advent of synthetic combinatorial library techniques has paved the way for the ever expanding diversity of compounds available for screening, increasing the chances of discovering novel molecules of known function and expanding fundamental knowledge of biological mechanisms and potentially the discovery of effective disease treatments.

Adding to a growing number of reviews on the field of combinatorial libraries (Scott, 1992; Houghten et al., 1992; Dooley et al., 1993; Houghten, 1993; Moos et al., 1993; Pavia et al., 1993; Scott and Craig, 1994; Sebestyen et al., 1993; Gallop et al., 1994; Gordon et al., 1994; Houghten, 1994) and the one-bead one-structure strategy (Lebl et al., 1995), this contribution will describe our recent developments in synthetic combinatorial libraries based on the one-bead one structure concept (Lam et al., 1991), with a focus on the design and synthesis of non-peptide libraries, the chemistry of releasable linkers as applied to solution phase assays and structure determination using codes for non-sequencable compounds.

2.2 One-Bead One-Structure Concept

The one-bead one-structure combinatorial library approach consists of three basic steps: (i) chemical synthesis yielding a library with a unique structure on each bead; (ii) screening the library using an on-bead binding assay or multi-step release assay; and (iii) structure determination for beads of interest. Library synthesis follows the mix and split method, first described by Furka (Furka et al., 1991), and illustrated in Figure 2.1. For the on-bead binding assay, beads containing active compounds are identified using an ELISA-type assay (Lam et al., 1991; Lam et al., 1993; Ohlmeyer et al., 1993; Lam and Lebl, 1992), or a fluorescent labeled probe (Chen et al., 1993; Needels et al., 1993; Meldal et al., 1994), and then physically segregated from the inactive compounds. Since not all projects are amenable to the on-bead assay, we have developed chemistries for stepwise release of compounds from library beads and extended the one-bead one-structure technique to solution phase assays (Lebl et al., 1993, Kocis et al., 1993; Salmon et al., 1993). The final step of the initial screen is structure determination of active

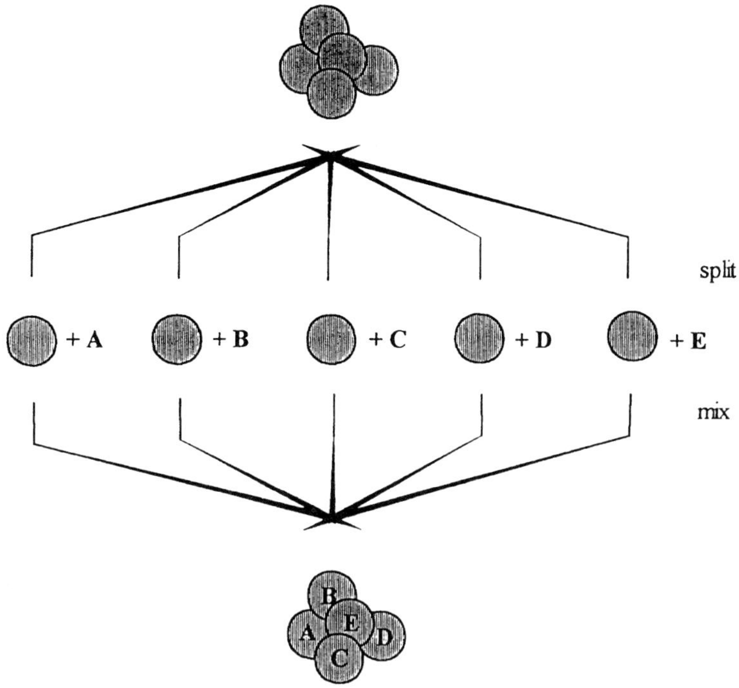

Figure 2.1: Schematic representation of Furka's split/mix method. Beads are equally split into a number (five in this Figure) of portions and in each portion a reaction with different building block (A, B, etc.) is performed. Then the beads are mixed and the process can be repeated.

compounds, using either direct methods, such as Edman degradation (Edman, 1950; Edman and Begg, 1967) and mass spectroscopy, or indirect coding approaches.

2.3 Design and Synthesis of Non-Peptide Libraries

The first synthetic combinatorial libraries were composed of peptides for obvious and pragmatic reasons. The chemistry used for synthesis on solid phase was well described (for review see e.g. (Atherton and Sheppard, 1989; Fields and Noble, 1990) and references cited therein), as were methods for structure determination with small amounts of material. Although peptides represent an important class of organic compound, numerous groups, including us, have embarked on an effort to create combinatorial libraries of non-peptidic small molecules (DeWitt et al., 1993; Zuckermann et al., 1992; Simon et al., 1992; Nikolaiev et al., 1993; Cho et al., 1993; Simon et al., 1994; Chen et al., 1994; Stankova et al., 1994; Lebl et al., 1994a; Lebl et al., 1994b, Krchňák et al. 1995a). The rational for this effort is two fold. First, the structural diversity of peptides is limited by the character of the peptide backbone. In addition, other classes of compounds have been shown to be structurally and chemically diverse and have characteristics not present in peptides that are important to drug candidates – such as oral bioavailability and resistance to protease degradation.

2.3.1 Diverse and Complex Libraries

To effectively create a large chemical diversity one needs reliable and high yield chemical reactions that can be performed on solid phase, as well as a sizable collection of building blocks that afford the possibility of different types of molecular interactions. Peptide libraries can be extremely complex, in that they contain large numbers of compounds of differing sequence – although these may be similar in structure. *N* amino acids used in each of x randomization steps create n^x structures. Using only 20 natural amino acids one can easily synthesize millions of peptides (e.g. all 3.2 million possible pentapeptides). However, while this is quite an achievement never before possible, mere numbers do not necessarily

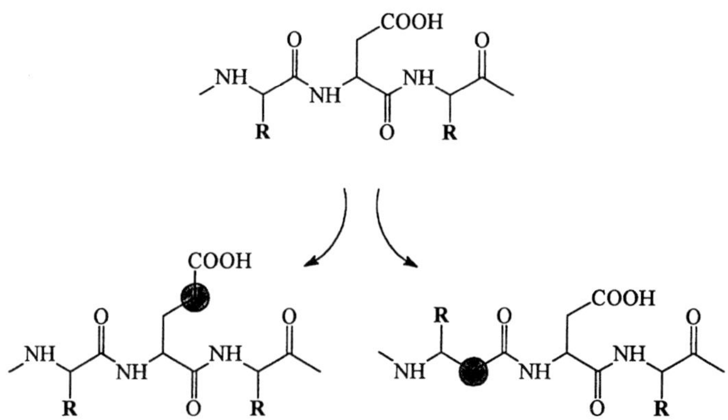

Figure 2.2: Changing the diversity of libraries by insertion of one methylene group.

increase the probability of discovering a desirable compound. Instead, the key to success in many instances may be the number of structurally unique compounds in a library; that is, the diversity within a library. Libraries of peptides do not represent a great diversity of structure, since the only changing parameter is the type of side-chain connected to the α-carbon of peptide backbone, and those side-chains can occupy only a predetermined conformational space. Combining L amino acids with D achieves greater diversity by enlarging the conformational space (Ramachandran plot (Ramachandran and Sassisekharan, 1968)), nevertheless, it is still quite limited. As an example, if a methylene group is inserted into the side chain, replacing an aspartyl residue with glutamyl (a typical scenario in peptide libraries), the number of components will double, but the diversity will not change substantially since it results in only the extension of one side chain. However, if the methylene group is inserted into the backbone, forming a β-amino acid, the complexity will also be doubled, but the diversity will be enhanced compared with the former case, since this insert will influence the relative spacing of the side chains. Figure 2.2 illustrates this point.

The differing characteristics of diverse and complex libraries can be used to one's advantage depending on the application. For example, a large number of building blocks of the same type (α-amino acids) coupled in the same format (peptide) will map a given conformational space very densely; however, the breadth of space covered by this library may be relatively small. If this space can accommodate crucial structural elements of compounds known to interact with the target molecule, a dense library may yield high affinity ligands. However, if the structural requirements for binding are not known, increasing the number of components in such a library will not necessarily enhance the probability of finding a hit. In this case, a more fruitful tactic may be to screen a diverse library that maps a large conformational space with lesser density. Hits identified from this library may not have the desired level of activity, but may be taken as a lead for further optimization. Subsequent libraries, containing a structural bias based on the initial lead can be designed providing maximum complexity in order to fine-tune the activity of the first hit.

2.3.2 Flexible and Rigid Libraries

In a simplified model, there are two primary determinants of whether a compound will bind specifically to a target molecule. The ligand should have the critical structural elements (functional groups) and present them in positions predetermined by the target. The first condition is readily met by selecting building blocks which represent all major sources of interactions known to play a significant role in molecular recognition. The second condition is closely related to the conformational space that is covered by individual library members. To enhance the probability that critical functional groups will be located properly without *a priori* knowing where they must be oriented, the library should cover a large conformational space. The one bead one structure concept presents a single chemical entity on each bead, but each individual is presented in a number of different conformations, depending on its flexibility. The higher the flexibility of structures, the larger the conformational space covered by individual members of a library. However, since there will be present many different conformers of one structure, the correct one (if present) will not be highly populated at any time. The trade off in this approach is that one may have to settle for relatively low affinity interactions as a result of an adverse entropy factor. Conversely, one may have a lower probability of identifying a high affinity interaction in screening libraries of more rigid structures. At present, there is no definitive means to quantify changes in the free energy of binding between the target molecule and ligand as a function of conformational freedom. An interim solution to this dilemma can be found by using the intuition and experience of the chemist, and knowledge of the target of interest. Future data should enable a more definitive answer regarding the optimal degree of flexibility. In the meantime, however, the speed with which libraries can be synthesized and screened enables one to try multiple approaches.

2.3.3 Examples of Libraries

Every combinatorial library is biased by the selection of building blocks, coupling chemistries and the linkers or "scaffolds" to which the building blocks are attached. The selection criteria are most likely based on a number of factors, including knowledge of the target molecule and availability of compatible build-

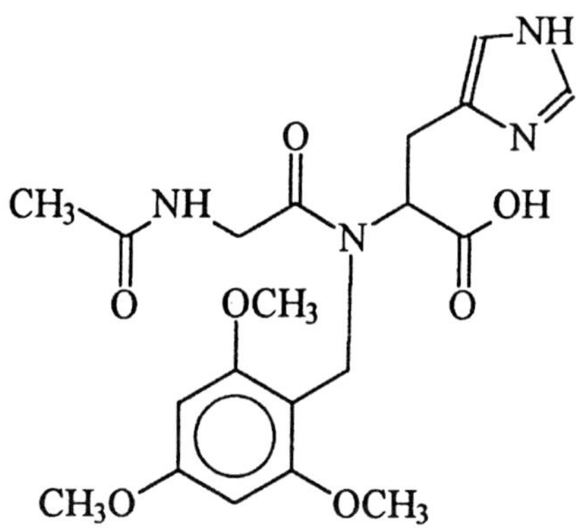

Figure 2.3: Structure of streptavidin binder from the library of alkylated and acylated amino acids.

ing blocks and coupling chemistries. The following is a list of various libraries we have synthesized and screened:

(i) Libraries of small, compact, and relatively rigid structures (e. g. N-acyl-N-alkyl amino acids (Stankova et al., 1994)).

(ii) Libraries based on a scaffold structure with variable rigidity (usually a multifunctional cyclic scaffold, e.g. a derivatized cyclopentane or cyclohexane ring, benzenetricarboxylic and diaminobenzoic acids (Lebl et al., 1994b; Kocis et al., 1994)).

(iii) Libraries based on a flexible scaffold that is built during the synthesis of the library and can be randomized (branched scaffold based on diamino acids, α,β,γ-library; Krchňák et al. 1995b).

To elaborate further on a representative non-peptidic small molecule library we have synthesized, we describe here the synthesis of a library of acylated and alkylated amino acids. Twenty α-amino acids were coupled to the resin beads. The amino protecting group was removed and the liberated amino group was reacted with a set of 20 aldehydes. The resulting Schiff base was protonated and reduced by sodium cyanoborohydride and the secondary amino group was acy-

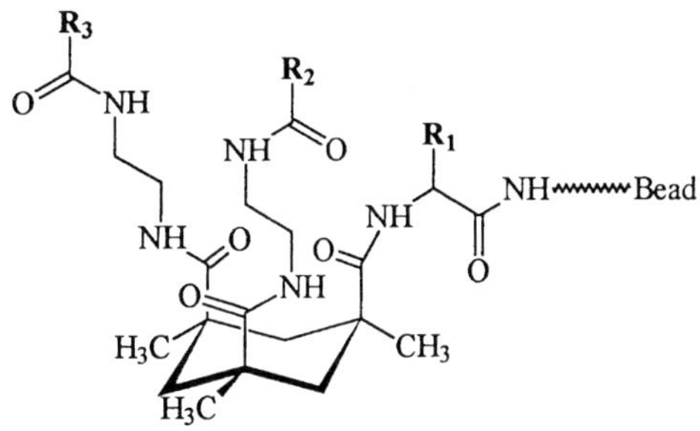

I

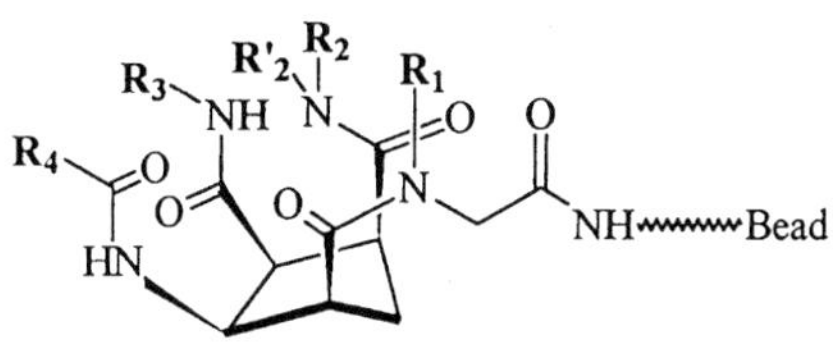

II

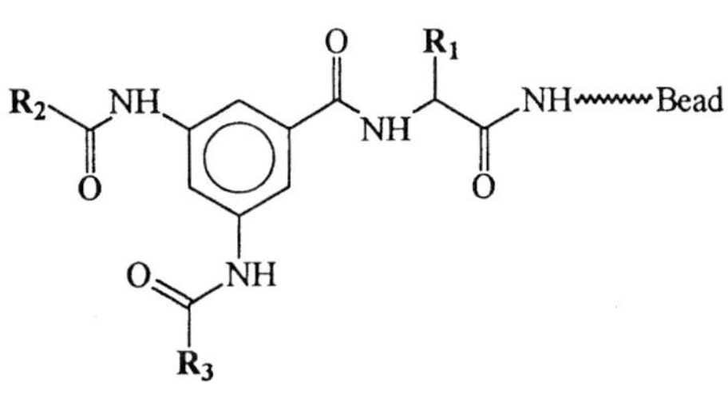

III

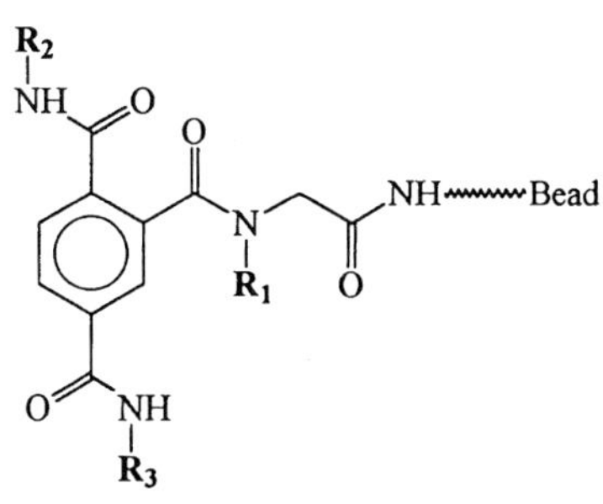

IV

Figure 2.4: Structure of scaffold based libraries.

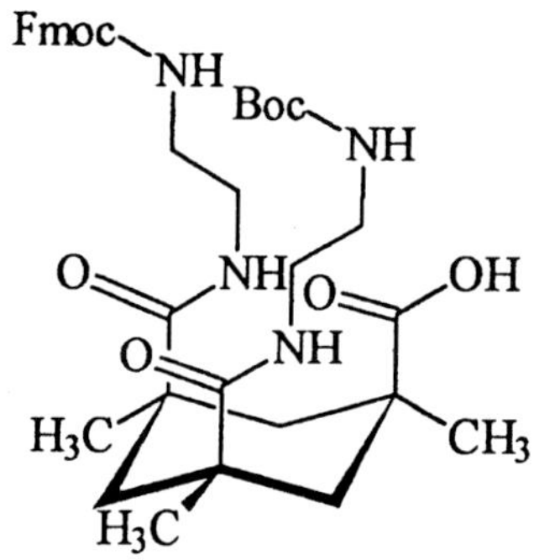

I

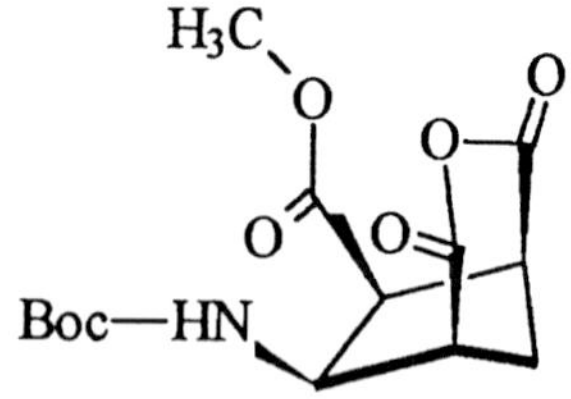

II

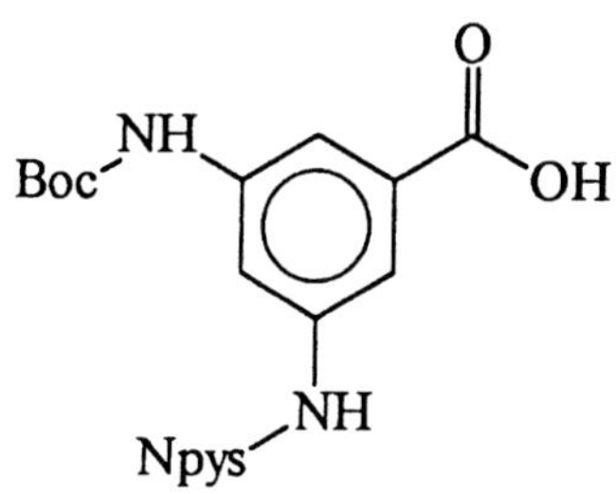

III

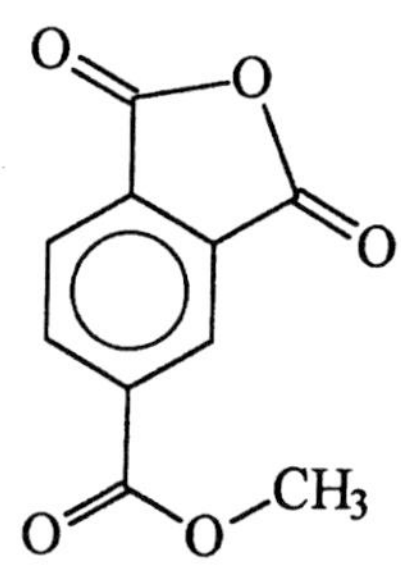

IV

Figure 2.5: Structure of protected scaffolds used in library synthesis. The following protecting group abbreviations were used: fluorenylmethyloxycarbonyl, Fmoc; t-butyloxycarbonyl, Boc; nitrophenylsulphenyl, Npys.

lated by 20 different carboxylic acids. We have developed a technique by which to identify the structure of these compounds using mass spectroscopy, avoiding the need to provide a sequencable coding tag (Stankova et al., 1994). Screening this library against streptavidin yielded three positive beads, each containing the same structure (see Figure 2.3). The compound was resynthesized and binding was confirmed in solution and found to be higher than that of previously identified peptide ligands (Lam et al., 1991; Lam and Lebl, 1992).

In order to constrain the presentation of a given set of functional groups, we have synthesized several libraries based on scaffolds with variable rigidity. Examples of scaffold based libraries are represented by four different cyclic skeletons (Figure 2.4): trimethylcyclohexanetricarboxylic acid (Kemp's triacid, structure I, Kocis et al. 1995), aminocyclopentanetricarboxylic acid (II, Patek et al. 1994), diaminobenzoic acid (III) and benzenetricarboxylic acid (IV). In all cases

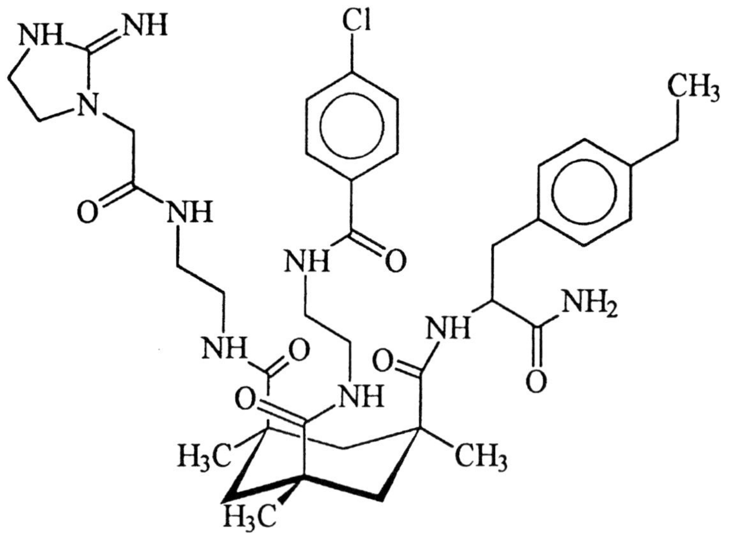

Figure 2.6: Structure of a thrombin inhibitor identified from a library based on a cyclohexane tricarboxylic acid scaffold.

suitably protected scaffolds have been synthesized (Figure 2.5) and coupled via a carboxyl group to an amino acid on the resin bead. The protecting groups are then removed one by one; each deprotection followed by coupling a set of building blocks – carboxylic acids when a liberated amino group is present on the resin, and amines when a free carboxyl group is available on the scaffold. The library based on the Kemp's triacid scaffold was tested against thrombin and five hits were found, the best of which (Figure 2.6) exhibited a Ki of 4µM (Kocis, P., et al., in preparation), higher than the classical active site peptide inhibitor fPRPG (20µM) which is the active site pharmacophore of Hirulog, currently in clinical trials.

As an example of a flexible library with randomized backbone, we synthesized a library with variable side chain spacing, having an α, β or γ amide bond between each subunit. The synthesis of this library will be outlined in the section describing coding.

2.4 Release Assay

The one-bead one-structure concept is inherently suitable for performing the on-bead binding assay. However, not all assays are amenable to on-bead screening, and for some important targets relevant assays must be performed in solution. In principle it is possible to distribute beads one by one, cleave the compound from each bead, test solutions and then recover those beads that released compounds displaying activity in given test and determine their structures. However, this technique is applicable in small complexity libraries only. To be able to screen libraries containing millions of different compounds we have developed a two step releasable assay. In the first step, beads are distributed into a 96 well format filtration plate, each well containing ca 500 beads, one third of the compound is released from each bead into solution and the mixture of compounds is tested. Beads from positive wells are re-distributed into individual wells and the second portion of the compound is released. Each bead corresponding to an active solution is recovered, and the compound (or coding tag) remaining on the bead is used for structure determination (Lebl et al., 1993; Kocis et al., 1993; Salmon et al., 1993). An alternative approach to performing the releasable assay, based on repeated partial release from one type of linker, was introduced recently by Jayawickreme et al. (Jayawickreme et al., 1994).

2.4.1 Chemistry of Releasable Linkers

Releasable libraries have been constructed according to the scheme depicted in Figure 2.7. Functional groups (amino groups) on the resin were branched to produce three independent branches, two of which were used for attaching the test compound. The third branch linked the compound used for identification, which can be the same compound or a tag encoding the structure of any non sequencable compound. Test compounds were attached to the releasable arms via an ester bond; however, the ester bond was cleaved by two unique mechanisms – entropically favored cyclization resulting in diketopiperazine (DKP) formation and alkaline hydrolysis. The release of a peptide based on DKP formation was described (Bray et al., 1993; Bray et al., 1994; Maeji et al., 1992; Bray et al., 1991a; Bray et al., 1991b; Maeji et al., 1990); however the DKP moiety was cleaved from

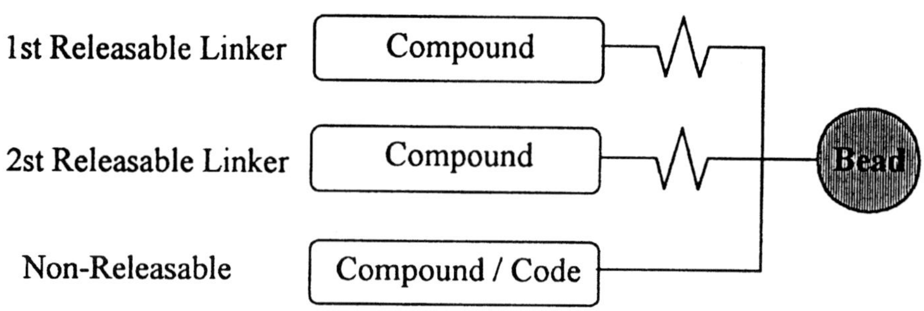

Figure 2.7: Design of double cleavable libraries.

the resin and stayed with the released compound. To release compounds with the same terminus in both stages and not containing DKP, we designed a "reverse" DKP linker, in which the DKP stays on the resin. Compounds are attached to the linker via an ester bond of Fmoc-Gly-NH-$(CH_2)_3$-OH (Fmoc-Gly-HOPA) and when released to the aqueous solution they contain an identical carboxy terminus, the hydroxypropylamide of glycine (Gly-HOPA).

The first generation of linkers was based on the Glu-Pro motif (Lebl et al., 1993) (Glu provides a side chain function and Pro enhances the tendency to cyclize). The recognition of iminodiacetic acid (Ida) as the α-amino acid involved in diketopiperazine formation allowed us to design novel double cleavable linkers. Iminodiacetic acid was found suitable for several reasons: (i) The imino group is in the α-position relative to the carboxyl groups; (ii) both carboxyl groups are chemically equivalent; (iii) as an N-substituted amino acid it is prone to cyclization via DKP formation with practically any other α-amino acid; (iv) it is not chiral; (v) it is inexpensive. In general, there are three variations of the Ida-based linker. They can be schematically depicted as dipeptides containing Aaa-Ida, Ida-Aaa, or Ida-Ida, where Aaa is any α-amino acid, preferably one that is prone to cyclization via DKP formation. We found that the position of Ida in such a dipeptide was not important. This is not true for the combination of Glu and Pro. The dipeptide Glu-Pro provides satisfactory kinetics of DKP formation unlike the dipeptide Pro-Glu, the cyclization of which takes more than 24 hours.

The dipeptide motif Ida-Ida was found particularly suitable for designing double cleavable linkers (Kocis et al., 1993). The Ida-Ida dipeptide is prone to DKP formation, it provides three carboxyl groups, one on the amino terminal Ida and two on the carboxy terminal. To construct the double cleavable linker, two carboxyl groups are needed for derivatization and subsequent synthesis of test compounds, one for attaching the linker to the resin beads. Two Fmoc-Gly-HOPA's are either coupled to both carboxyls of the carboxy terminal Ida (Figure 2.8, linker I), or each Ida bears one Fmoc-Gly-HOPA (linker II). In either case there is one free carboxyl group that serves for connecting the linker to the solid support, e.g. via Lys which provides one extra amino group for a third, nonreleas-able copy of the compound or the code. The chemistry of both releases is shown on Figure 2.9.

Using peptide libraries built on double cleavable linker, we have identified ligands for the anti-β-endorphin antibody and the glycoprotein IIb/IIIa receptor (Salmon et al., 1993).

The released compounds from both double releasable linkers I and II contained the Gly-HOPA. Since it may be desirable in some instances to release a compound without the Gly-HOPA, but having a free carboxyl group instead, we designed a modified linker that incorporates an additional ester linkage (Figure 2.10). The appended ester bond is introduced into the linker by attaching a hydroxy acid (e.g. 3-hydroxypropylamide of glutaric acid) to both arms of the linker. During the first release at pH8 the DKP is formed and compound with Gly-HOPA is released to the solution. The beads are then separated from the

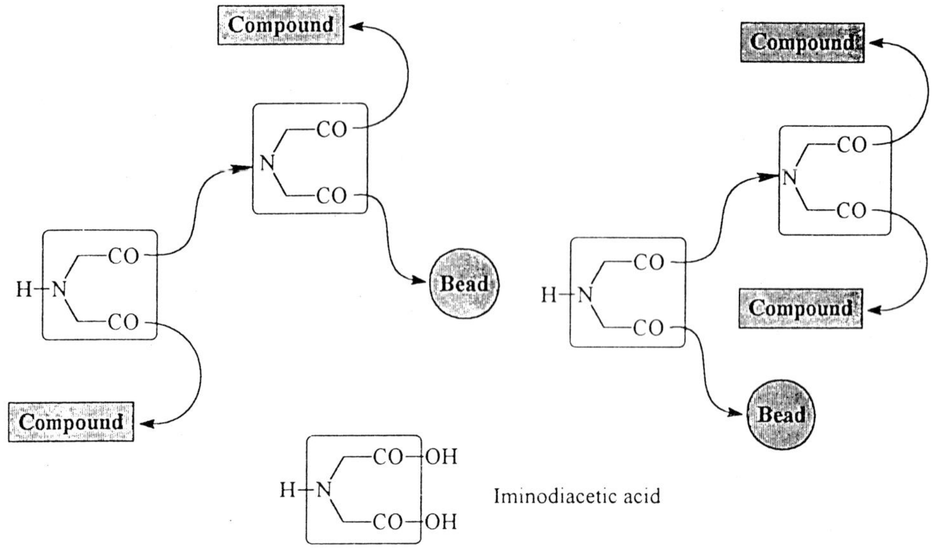

Figure 2.8: Scheme of two imminodiacetic acid based linkers.

solution by filtration, the pHis brought to ca 13 using NaOH, and after incubation to permit ester hydrolysis (typically ca 30 min) the solution is adjusted to physiologic pHfor biological screening. The second release is performed using NaOH as previously described and yields the desired compound with a free carboxyl group.

2.4.2 Sensitivity of the Assay

The on-bead binding assay is very sensitive, due to the existence of a high local concentration of test compound on the surface area of the resin bead. One gram of TentaGel resin (polyethylene glycol grafted polystyrene crosslinked with 1% divinylbenzene) of average size 130 µm and with a substitution at 0.2 mmol of amino groups per gram of resin contains ca 1 million beads. A simple calculation reveals that the local concentration of test compound at the surface area of each bead is approximately 20 mM. However, if the test compound from the same bead is released to solution, the final concentration is approximately 2 µM, assuming a 100 µl assay volume and quantitative release of test compound.

This number is critical from the assay point of view, since the weakest binder that can be detected will have an IC_{50} in the micromolar range. To detect weaker interactions, the volume of assay can be decreased (we have been able to decrease the vol-

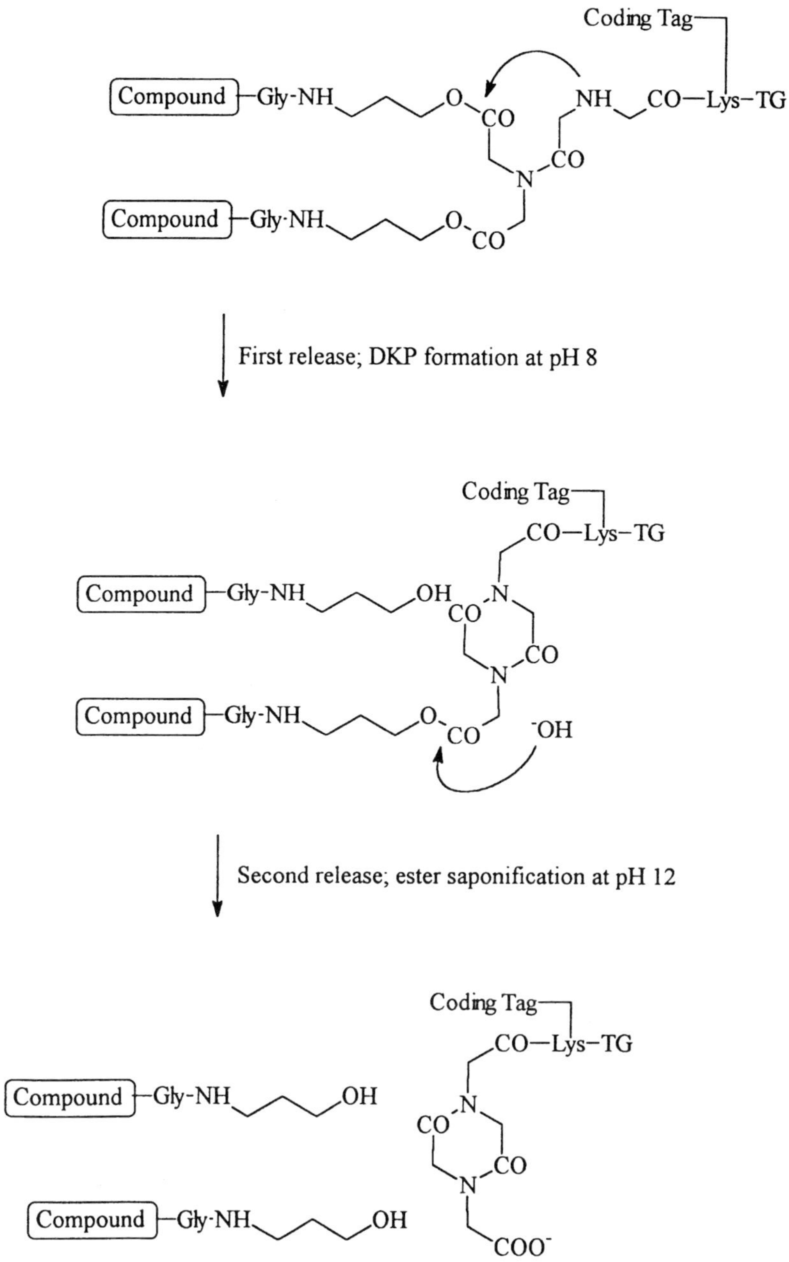

Figure 2.9: Chemistry of two stage release of a test compound from the iminodiacetic acid based linker.

ume three-fold). Alternatively, one can increase the amount of compound released from each bead. This can be achieved in three different ways: (i) increase the substitution of beads, maintaining the same size of beads, (ii) increasing the size of the beads, or (iii) using an alternative solid support, e.g. cotton thread or plastic membrane, that can be cut into pieces of almost any size, modulating the

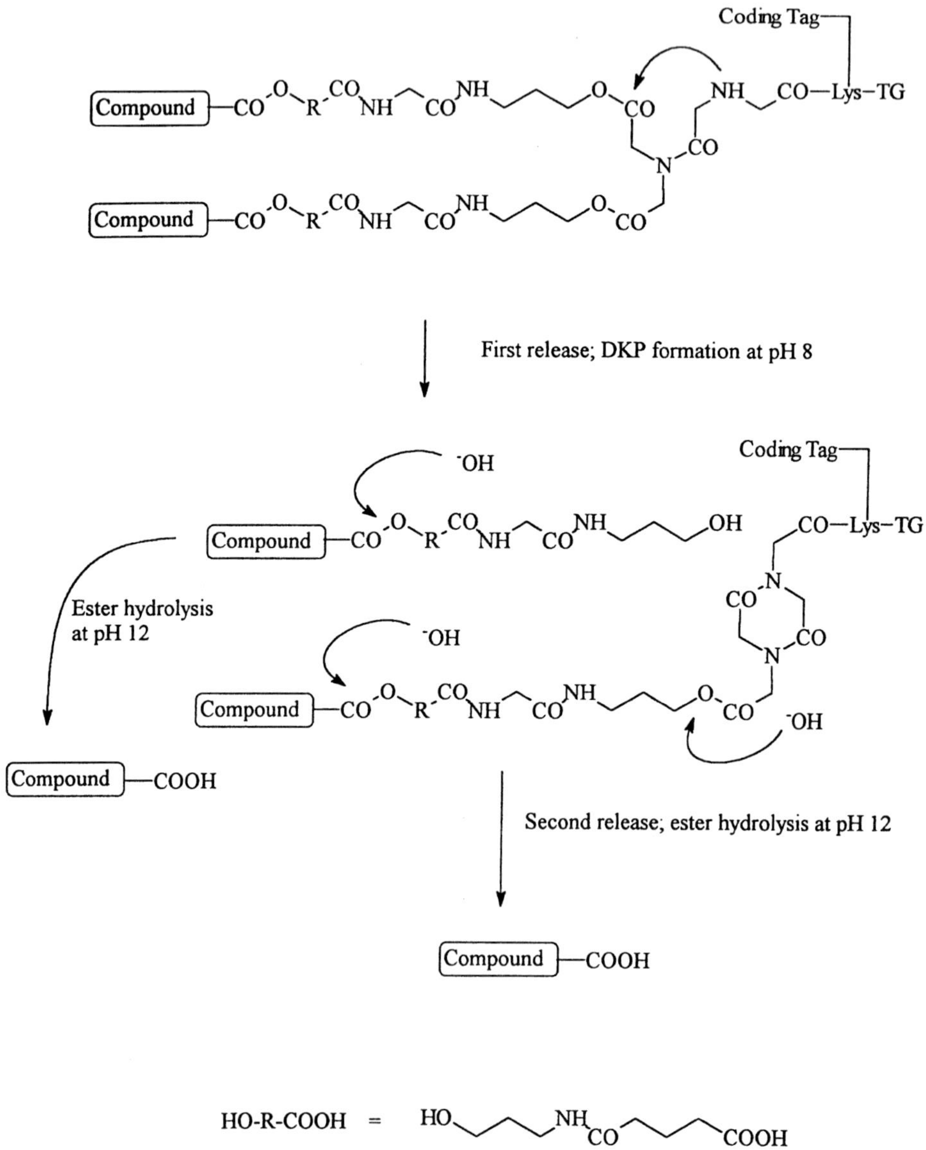

Figure 2.10: Chemistry of two stage release of a test compound with free carboxyl group.

releasable quantity by length or area of support. We have increased substitution up to 1 mmol of amino groups per gram of resin with the same size of beads with good success, but found that any further increase in substitution leads to slow reaction rates and incomplete coupling. We have used custom made 220 μm beads constructed of polystyrene grafted with polyethylene glycol that bring the release to the nmol range. Finally, synthesis on cotton thread or fabric is feasible (Eichler et al., 1991; Lebl et al., 1992; Rinnova et al., 1993), although difficult to handle in a library format.

2.5 Structure Determination

The final step in the one-bead one-structure approach is structure identification for selected beads. A limiting factor in this respect is the amount of compound available for analysis, typically in the range of 100 pmol. This is not sufficient to apply classical structure determination techniques (elemental analysis, spectroscopic methods, etc.). In the case of peptide libraries the problem can be solved easily since there is an established technique for the determination of the sequence of peptides, Edman degradation (Edman, 1950; Edman and Begg, 1967). This method has been automated and its sensitivity (2–3 pmol in our hands) is well below the amount available on one bead.

Determination of the structure of organic molecules in diverse combinatorial libraries requires alternative methods. We have applied two different techniques: direct structure determination using mass spectroscopy and an indirect coding method.

2.5.1 Mass Spectroscopy

There is enough material on one resin bead for analysis by mass spectroscopy. Due to the large number of individual components in one library, determination of molecular weight is not sufficient information to resolve the structure. Therefore we have applied MS/MS techniques, which provide information regarding fragmentation. Since the molecular weight of all building blocks is known, in the case of informative fragmentation it is in principle possible to recall the composition of individual library compounds. The success of this method depends on knowing the fragmentation pattern and we have been able to use this technique in a number of cases, including the N-alkyl-N-acylamino acid library (Stankova et al., 1994), the Kemp's triacid library, and libraries based on side chain substituted diamino acids assembled into a peptide chain. We have also applied deuterium exchange of labile protons to increase the reliability of structure determination (Sepetov et al., 1993).

2.5.2 Coding Principle

If direct structure determination is not feasible we apply, similarly as the others, the coding strategy (Ohlmeyer et al., 1993; Needels et al., 1993; Nikolaiev et al., 1993; Brenner and Lerner, 1992; Nielsen et al., 1993; Kerr et al., 1993), just as mother nature has been doing for ages by coding of protein sequences with nucleic acids. The chemistry of encoded library synthesis is more difficult and time consuming, since there must be two compatible chemistries available to build a screening arm and a coding arm at the same time. In the case of peptides as coding tags, each chemical reaction in a randomization step is coded by a corresponding amino acid (Figure 2.11). The structure of the resulting peptide code is resolved using Edman degradation (Edman, 1950; Edman and Begg, 1967). If a different set of coding amino acids is used in each randomization step, each amino acid codes for not only the building block, but also the randomization step. However, taken individually, the number of different amino acids would be too high (e.g. four randomizations, each performed with 45 building blocks would require 180 amino acids). To reduce the number of coding amino acids we have applied a binary code, in which each chemical reaction is coded by a doublet of amino acids (Sepetov et al., in preparation; binary coding using different chemistry has recently been described (Ohlmeyer et al., 1993)). In this arrangement n amino acids can code for $n(n-1)/2$ building blocks, substantially reducing the necessary number of coding amino acids. In addition, since both the encoded building block and the round of synthesis can be determined, it is possible to design the structure of the coding tag to permit cleavage of all the doublets simultaneously, substantially reducing the time for structure determination.

To have a battery of α-amino acids with desirable properties (suitable reten ion time of their phenylthiohydantoins on HPLC and compatibility with chemistries applied during library synthesis) we acylated the side chain amino group of

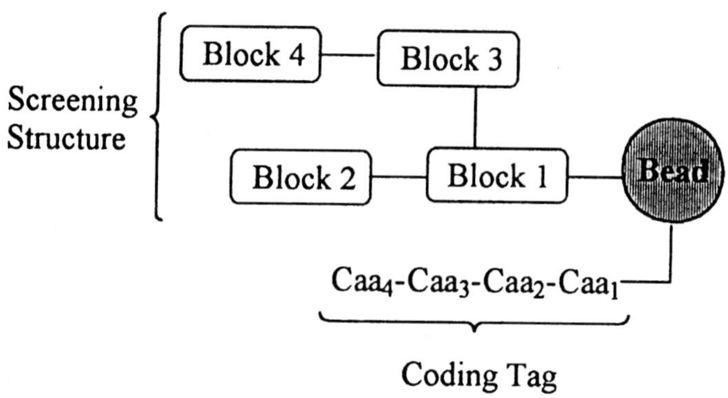

Figure 2.11: Principle of coding

diamino acids (lysine, ornithine, diaminobutyric, and diaminopropionic acids), evaluated retention time of their phenylthiohydantoins, and selected those that provided baseline separation.

2.5.3 Alternative Coding Techniques

When the library is synthesized in a releasable format, all components of the library are tested in solution, separated from the bead of origin. However, when the on-bead binding assay is employed, the coding structure may interact with the target molecule. There are at least three possibilities to prevent unwanted interaction of the coding structure: (i) present the coding structure in very low concentration, decreasing the probability of interaction; (ii) make the code inaccessible to the target molecule; (iii) code using a mixture of coding structures rather than by only one coding compound.

The first method can be achieved using an oligonucleotide sequence as the code (Needels et al., 1993; Brenner and Lerner, 1992; Nielsen et al., 1993). The high sensitivity of this method results from the use of PCR amplification. The relative abundance of coding tag can thus be minimized and the probability of interaction is diminished. An alternative and elegant scheme was recently described by Ohlmeyer et al. (Ohlmeyer et al., 1993), in which high sensitivity was gained by using halogenated aromatic acids analyzed by gas chromatography.

We have developed the second strategy, based on physical separation of the surface of the bead from its interior (Vagner et al., 1994a; Vagner et al., 1994b). TentaGel resin used for this concept is not penetrable by high molecular weight compounds, therefore the interaction with the macromolecular target occurs only on the surface. Several methods to separate reactive functional groups (amino groups) on the surface from those located inside beads were tested. The most promising results were obtained by enzymatic shaving (Figure 2.12). An enzyme

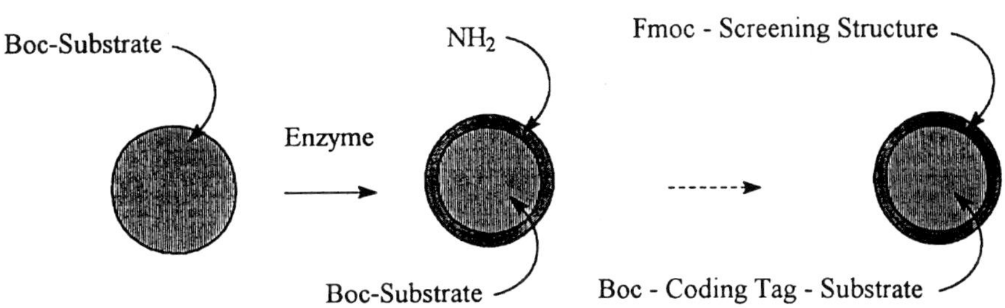

Figure 2.12: Enzymatic shaving of beads.

substrate, Boc-Phe-Gly, was synthesized on all available amino groups and then cleaved by chymotrypsin. Since the enzyme cannot penetrate inside the bead, only the substrate on the surface was cleaved (shaved). To show that the surface was efficiently depleted, we synthesized the streptavidin ligand LHPQF and the thrombin inhibitor fPRPG inside separate batches of beads. Neither ligand was detected by the corresponding assay, although both peptides were successfully sequenced. Both peptides were also synthesized on the shaved surface of beads and were then detectable in the corresponding assay.

In the third technique coding is not achieved by a single structure, but rather by a set of different structures having one common feature, thereby providing unambiguous identification. We explored three different strategies for design and synthesis of the coding arm. In the first approach (Figure 2.13, panel B), each chemical reaction performed on the screening arm is followed by deprotection of the amino group on the coding arm, coupling a mixture of two coding amino acids, protected so that the protecting group can be removed after the next chemical reaction on the screening arm is complete. The next mixture of two coding amino acids are then coupled to the amino groups of the previous coding amino acids. After finishing the synthesis, the coding arm represents a mixture of peptides, in the case of four coded chemical reactions the coding structures comprise 16 different tetrapeptides. Since four cycles of Edman degradation (Edman, 1950; Edman and Begg, 1967) are needed to determine the structure, the same set of coding amino acids can be used in each step, even though they can code for different reactions or building blocks.

Another approach uses only a portion of available amino groups for attaching the coding amino acids (Figure 2.13, panel A). After liberating the amino groups on the coding arm, approximately half of the equimolar amount of mixture of the two coding amino acids is coupled. The remaining 50 % of free amino groups on the coding arm is then reacted with a protected diamino acid. In this way the number of remaining amino groups is doubled, reaching the original level of substitution, since only 50 % of amino groups were doubled. In this arrangement a different set of acids must be used for each coded step because the sequencing will reveal all coding amino acids in one step of Edman degradation and the coding amino acids have to bear not only the information about the chemistry but also code for the step number.

In the last version of coding strategy all coding amino acids are linked to one amino group (Figure 2.13, panel C). The protecting group from the coding arm is removed and the first doublet of coding amino acids is coupled to a portion (e.g. 25 % in the case of four coding steps) of available amino groups. The rest (75 %) of the amino groups are blocked by a suitable protecting group, that can be removed before the second coding step. The limiting factor in this scheme is the amount of each coding amino acid needed to see a reasonable signal upon sequencing.

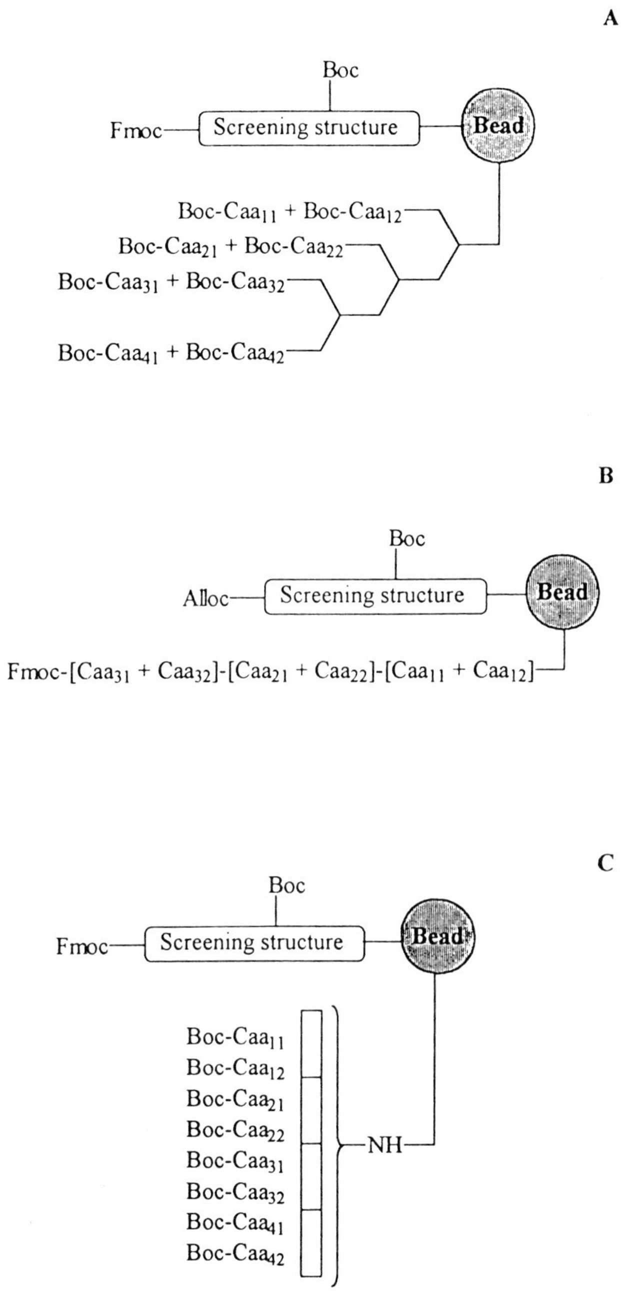

Figure 2.13: Three strategies for design of the digital coding tag. Orthogonal combination of protecting groups was used for the construction of the screening structure (Fmoc and Boc in the panel A and C, Alloc and Boc in the panel B) and coding tag (Boc in panels A and C, Fmoc in panel B).

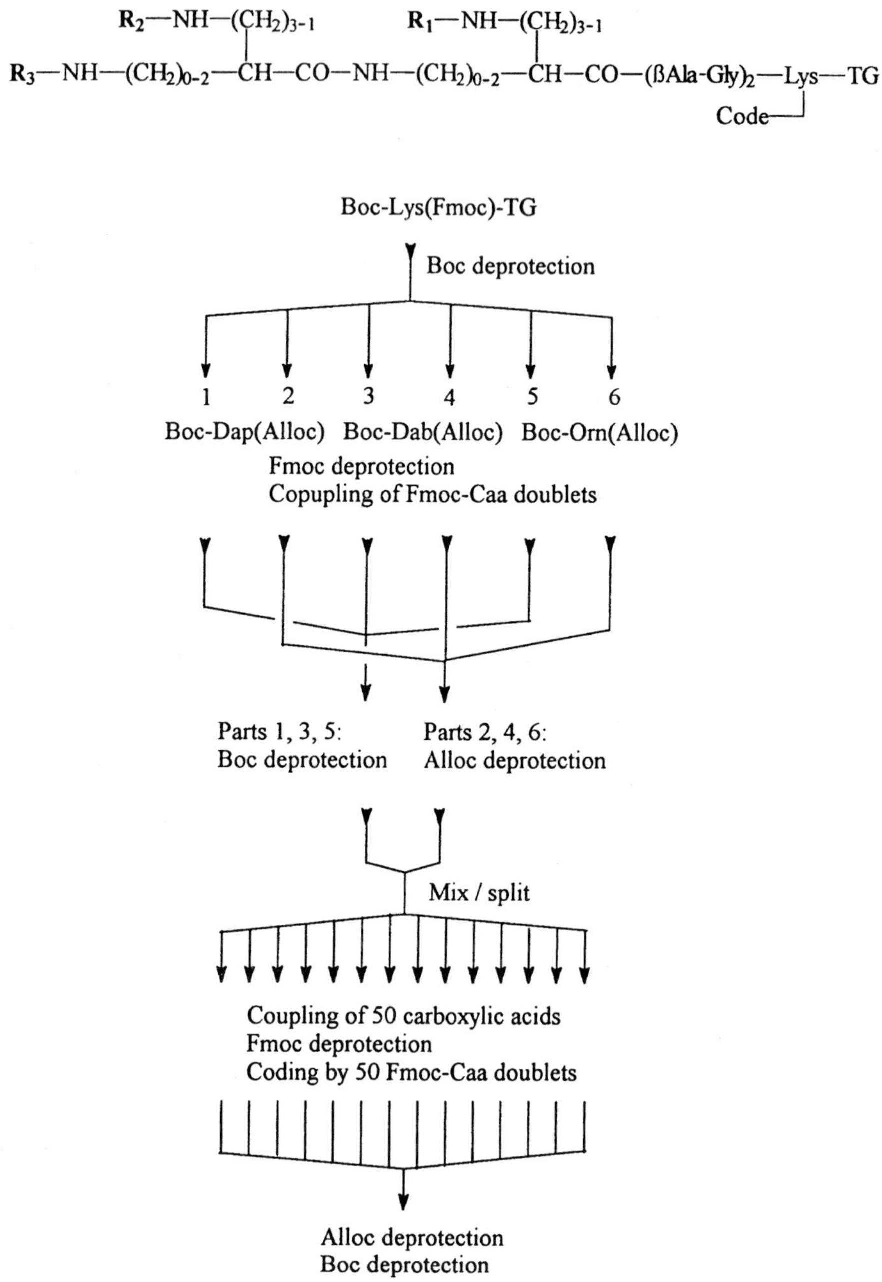

Figure 2.14: Synthetic scheme of coded library having variable side-chain spacing.

Coding in a more general sense tracks the history of synthesis performed on the resin beads. The coding tag does not necessarily need to include that which was coupled to the resin, but can be used to store, for example, the information what was removed from the resin. This type of coding is illustrated by the synthesis of a library with variable side-chain spacing (α,β,γ-library; Krchňák et al. 1995b). In this library (Figure 2.14) we attached a set of building blocks (carboxylic acids) to one amino group of a diamino acid (diaminopropionic acid, diaminobutyric acid, and ornithine) and used the second amino group for the extension of the backbone by coupling another diamino acid. Half of the library had the carboxylic acid coupled via the α-amino group, the second half had coupled the acids to the side chain amino group.

Synthesis was performed using N^{α} Boc, N^{ω} Alloc protected diamino acids. Coding amino acids were protected by an Fmoc group. Boc-Lys(Fmoc) was coupled to the resin and its α-amino group was used to build the screening arm. The sequencing arm was synthesized on the ε-amino group. The linker βAla-Gly-βAla-Gly was assembled on the α-amino group to separate the screening arm. After removing the Boc protecting group the resin was divided into 6 parts and diaminopropionic acid was coupled to parts 1 and 2, diaminobutyric acid to parts 3 and 4 and ornithine to the remaining two parts. The Fmoc group from the coding arm was then cleaved and 6 different doublets of coding amino acids were coupled to the coding arm. In the next step parts 1, 3, and 5 were combined and the Boc group was removed. Parts 2, 4, and 6 were also combined and the Alloc group was removed. Both parts were combined, the beads were thoroughly mixed and distributed into 50 parts and a set of 50 carboxylic acids was coupled, followed by coding using 50 doublets of coding amino acids (requiring only 11 coding subunits). After finishing the library synthesis the entire coding arm consisted of a mixture of heptapeptides that coded for four randomizations of carboxylic acids (sequencing cycles No 1, 2, 4, and 6) and the type of diamino acid and, at the same time, coded for the protecting group which was removed from the diamino acid (cycles No 3, 5, and 7).

2.6 Conclusion

This contribution has focused exclusively on the one-bead one-structure technique applied at Selectide, illustrating developments in three areas that we consider critical to the application of the process. Several examples were documented in which interesting results have been obtained. The structural and chemical diversity represented in combinatorial libraries through the use of building blocks other than amino acids and the use of scaffolds rather than a peptide backbone were given. Notable are the inhibitors of thrombin identified from the

Kemp's library. Not only was the most active compound higher affinity than the classical active site peptide inhibitor fPRPG, it is not a linear molecule. This is a clear validation of success in increasing library chemical and structural diversity in the Selectide process. The second point, addressing aspects relating to the release and screening of compounds in solution, focused on the need to be able to apply the Selectide Process to all target not just select few purified proteins. The third point focused on methods which have been developed to identify the composition of active compound in the case where they are not peptides. These examples serve to demonstrate that our approach, along with those of several other groups represent an increasingly important set of tools which can be applied to the discovery of new pharmaceuticals and to the increased understanding of molecular bioscience.

References

Atherton, E. and Sheppard, R.C. (1989) Solid Phase Peptide Synthesis. (IRL Press: Oxford).

Bray, A. M., Maeji, J. N., Jhingran, A. G., and Valerio, R. M. (1991). Gas Phase Cleavage of Peptides from a Solid Support with Ammonia Vapour. Application in Simultaneous Multiple Peptide Synthesis. Tetrahedron Lett. *32,* 6163–6166.

Bray, A. M., Maeji, J. N., Valerio, R. M., Campbell, R. A., and Geysen, M.H. (1991). Direct Cleavage of Peptides from a Solid Support into Aqueous Buffer. Application in Simultaneous Multiple Peptide Synthesis. J. Org. Chem. *56,* 6659–6666.

Bray, A.M., Valerio, R.M., and Maeji, N.J. (1993). Cleavage of Resin-Bound Peptide Esters with Ammonia Vapour -Simultaneous Multiple Synthesis of Peptide Amides. Tetrahedron Lett. *34,* 4411–4414.

Bray, A.M., Jhingran, A.G., Valerio, R.M., and Maeji, N.J. (1994). Simultaneous Multiple Synthesis of Peptide Amides by the Multipin Method – Application of Vapor-Phase Ammonolysis. J. Org. Chem. *59,* 2197–2203.

Brenner, S., and Lerner, R.A. (1992). Encoded combinatorial chemistry. Proc. Natl. Acad. Sci. USA *89,* 5381–5383.

Bunin, B.A., Plunkett, M.J., and Ellman, J.A. (1994). The combinatorial synthesis and chemical and biological evaluation of 1,4-benzodiazepine library. Proc. Natl. Acad. Sci. USA *91,* 4708–4712.

Cass, R., Dreyer, M.L., Giebel, L.B., Hudson, D., Johnson, C.R., Ross, M.J., Schaeck, J., and Shoemaker, K.R. (1994). Pilot, a new peptide lead optimization technique and its application as a general library method. In: Peptides, Proc. 13th APS, R.S. Hodges and J.A. Smith, eds. (ESCOM: Leiden), pp. 975–977.

Chen, J.K., Lane, W.S., Brauer, A.W., Tanaka, A., and Schreiber, S.L. (1993). Biased combinatorial libraries: Novel ligands for the SH3 domain of phosphatidylinositol 3-kinase. J. Am. Chem. Soc. *115,* 12591–12592.

Chen, C., Randall, A.L.A., Miller, B.R., Jones, D.A., and Kurth, M.J. (1994). Analogous Organic Synthesis of Small-Compound Libraries: Validation of Combinatorial Chemistry in Small Molecule Synthesis. J. Am. Chem. Soc. *116,* 2661–2662.

Cho, C.Y., Moran, E.J., Cherry, S.R., Stephans, J.C., Fodor, S.P.A., Adams, C.L., Sundaram, A., Jacobs, J.W., and Schultz, P.G. (1993). An unnatural biopolymer. Science *261,* 1303–1305.

Cull, M.G., Miller, J.F., and Schatz, P.J. (1992). Screening for receptor ligands using large libraries of peptides linked to the C terminus of the lac repressor. Proc. Natl. Acad. Sci. USA *89*, 1865–1869.

Cwirla, S.E., Peters, E.A., Barrett, R.W., and Dower, W.J. (1990). Peptides on phage: A vast library of peptides for identifying ligands. Proc. Natl. Acad. Sci. USA 87, 6378–6382.

Devlin, J.J., Panganiban, L.C., and Devlin, P.E. (1990). Random peptide libraries: A source of specific protein binding molecules. Science *249*, 404–406.

DeWitt, S.H., Kiely, J.S., Stankovic, C.J., Schroeder, M.C., Reynolds Cody, D.M., and Pavia, M.R. (1993). "Diversomers": An approach to nonpeptide, nonoligomeric chemical diversity. Proc. Natl. Acad. Sci. USA *90*, 6909–6913.

Dooley, C.T., and Houghten, R.A. (1993). The Use of Positional Scanning Synthetic Peptide Combinatorial Libraries for the Rapid Determination of Opioid Receptor Ligands. Life Sci. *52*, 1509–1517.

Edman, P. (1950). Method for determination of the amino acid sequence in peptides. Acta Chem. Scand. *4*, 283–293.

Edman, P., and Begg, G. (1967). A protein sequenator. Automated equipment for sequence determination. Eur. J. Biochem. *1*, 80–91.

Eichler, J., Bienert, M., Stierandova, A., and Lebl, M. (1991). Evaluation of cotton as a carrier for solid phase peptide synthesis. Pept. Res. *4*, 296–307.

Felici, F., Castagnoli, L., Musacchio, A., Jappelli, R., and Cesareni, G. (1991). Selection of antibody ligands from a large library of oligopeptides expressed on a multivalent exposition vector. J. Mol. Biol. *222*, 301–310.

Fields, G.B., and Noble, R.L. (1990). Solid phase peptide synthesis utilizing 9-fluorenylmethyloxycarbonyl amino acids. Int. J. Peptide Protein Res. *35*, 161–214.

Fodor, S.P.A., Leighton, R.J., Pirrung, M.C., Stryer, L., Tsai, L.A., and Solas, D. (1991). Light-directed, spatially addressable parallel chemical synthesis. Science *251*, *767–773*.

Frank, R., and Döring, R. (1988). Simultaneous multiple peptide synthesis under continuous flow conditions on cellulose paper discs as segmental solid supports. Tetrahedron Lett. *44*, 6031–6040.

Furka, A., Sebestyen, F., Asgedom, M., Dibo, G. (1988a). Cornucopia of peptides by synthesis. (Poster Presentation). 14th International Congress of Biochemistry (Prague) Vol. 5., p. 47.

Furka, A., Sebestyen, F., Asgedom, M., Dibo, G. (1988b). More peptides by less labour. (-Poster Presentation). 10th International Symposium on Medicinal Chemistry (Budapest) p. 288.

Furka, A., Sebestyen, F., Asgedom, M., and Dibo, G. (1991). General method for rapid synthesis of multicomponent peptide mixtures. Int. J. Peptide Protein Res. *37*, 487–493.

Gallop, M.A., Barrett, R.W., Dower, W.J., Fodor, S.P.A., and Gordon, E.M. (1994). Applications of Combinatorial Technologies to Drug Discovery.Background and Peptide Combinatorial Libraries. J. Med. Chem. *37*, 1233–1251.

Geysen, M.H., Meloen, R.H., and Barteling, S.J. (1984). Use of peptide synthesis to probe viral antigens for epitopes to a resolution of a single amino acid. Proc. Natl. Acad. Sci. USA *81*, 3998–4002.Geysen, M.H., Rodda, S.J., and Mason, T.J. (1986). A priori delineation of a peptide which mimics a discontinuous antigenic determinant. Mol. Immunol. *102*, 709–715.

Gordon, E.M., Barrett, R.W., Dower, W.J., Fodor, S.P.A., and Gallop, M.A. (1994). Applications of combinatorial technologies to drug discovery. Combinatorial organic synthesis, library screening strategies, and future directions. J. Med. Chem. *37*, 1385–1401.

Houghten, R.A. (1985). General method for the rapid solid-phase synthesis of large numbers of peptides: Specificity of antigen antibody interaction at the level of individual amino acids. Proc. Natl. Acad. Sci. USA *82*, 5131–5135.

Houghten, R.A., Pinilla, C., Blondelle, S.E., Appel, J.R., Dooley, C.T., and Cuervo, J.H. (1991). Generation and use of synthetic peptide combinatorial libraries for basic research and drug discovery. Nature *354*, 84–86.

Houghten, R.A., Appel, J.R., Blondelle, S.E., Cuervo, J.H., Dooley, C.T., and Pinilla, C. (1992). The use of synthetic peptide combinatorial libraries for the identification of bioactive peptides. BioTechniques *13*, 412–421.

Houghten, R.A. (1993). Peptide Libraries – Criteria and Trends. Trends in Genetics *9*, 235–239.

Houghten, R.A. (1994). Soluble combinatorial libraries: extending the range and repetoire of chemical diversity. Methods, A Companion to Methods in Enzymology, *6*, 354–360.

Jayawickreme, C.K., Graminski, G.F., Quillan, J.M., and Lerner, M.R. (1994). Creation and Functional Screening of a Multi-Use Peptide Library. Proc. Natl. Acad. Sci. USA *91*, 1614–1618.

Kerr, J.M., Banville, S.C., and Zuckermann, R.N. (1993). Encoded combinatorial peptide libraries containing non-natural amino acids. J. Am. Chem. Soc. *115*, 2529–2531.

Kocis, P., Krchňák, V., and Lebl, M. (1993). Symmetrical structure allowing the selective multiple release of a defined quantity of peptide from a single bead of polymeric support. Tetrahedron Lett. *34*, 7251–7252.

Kocis, P., Ferguson, R., Krchňák, V., Wade, S., Lebl, M. (1994). Synthesis of nonpeptide libraries for study of host-guest interaction. (Poster Presentation). 8th International Sumposium on Molecular Recognition and Inclusion (Otawa)

Kocis, P., Issakova, O., Sepetov, N.F., and Lebl, M. (1995). Kemp's triacid scaffolding for synthesis of combinatorial nonpeptide uncoded libraries. Tetrahedron Lett. in press.

Kramer, A., Volkmer-Engert, R., Malin, R., Reineke, U., and Schneider-Mergener, J. (1993). Simultaneous synthesis of peptide libraries on single resin and continuous cellulose membrane supports: Examples for the identification of protein, metal and DNA binding peptide mixtures. Pept. Res. *6*, 314–319.

Krchňák, V., Flegelová, Z., Weichsel, A., and Lebl, M. (1995a). Polymer-Supported Mitsunobu Ether Formation and its Use in Combinatorial Chemistry. Tetrahedron Lett. in press.

Krchňák, V., Weichsel, A.S., Cabel, D., and Lebl, M. (1995b). Linear Flexible Library with Variable Side-chain Spacing: the α, β, γ-Library. Pept. Res. in press.

Krchňák, V., and Vagner, J. (1990). Color-monitored solid-phase multiple peptide synthesis under low-pressure continuous flow conditions. Pept. Res. *3*, 182–193.

Lam, K.S., Salmon, S.E., Hersh, E.M., Hruby, V.J., Kazmierski, W.M., and Knapp, R.J. (1991). A new type of synthetic peptide library for identifying ligand-binding activity. Nature *354*, 82–84.

Lam, K.S., and Lebl, M. (1992). Streptavidin and Avidin Recognize peptide ligands with different motifs. Immunomethods *1*, 11–15.

Lam, K.S., Hruby, V.J., Lebl, M., Knapp, R.J., Kazmierski, W.M., Hersh, E.M., and Salmon, S.E. (1993). The chemical synthesis of large random peptide libraries and their use for the discovery of ligands for macromolecular acceptors. Bioorg. Med. Chem. Lett. *3*, 419–424.Lebl, M., Stierandova, A., Eichler J, Patek, M., Pokorny V, Jehnicka J, Mudra, P., Zenisek, K., and Kalousek J. (1992). An automated multiple solid phase peptide synthesizer utilizing cotton as a carrier. In: Innovation and Perspectives in Solid Phase Synthesis, R. Epton, ed. (Intercept: Andover), pp. 251–257.

Lebl, M., Patek, M., Kocis, P., Krchňák, V., Hruby, V.J., Salmon, S.E., and Lam, K.S. (1993). Multiple release of equimolar amounts of peptides from a polymeric carrier using orthogonal linkage cleavage chemistry. Int. J. Peptide Protein Res. *41*, 201–203.

Lebl, M., Krchňák, V., Safar, P., Stierandova, A., Sepetov, N.F., Kocis, P., and Lam, K.S. (1994a). Construction and screening of libraries of peptide and non-peptide structures. In: Techniques in Protein Chemistry, J.W. Crabb, ed. (Academic Press: San Diego), pp. 541–548.

Lebl, M., Krchňák, V., Stierandova, A., Safar, P., Kocis, P., Nikolaiev, V., Sepetov, N.F., Ferguson, R., Seligmann, B., Lam, K.S., and Salmon, S.E. (1994b). Nonsequenceable and/or nonpeptide libraries. In: Peptides, Proc.13th APS, R.S. Hodges and J.A. Smith, eds. (ESCOM: Leiden), pp. 1007–1008.

Lebl, M., Krchňák, V., Sepetov, N.F., Seligmann, B., Strop, P., Felder, S., and Lam, K.S. (1995). One-Bead One-Structure Cominatorial Libraries. Biopolymers *37*, 177–198.

Maeji, N.J., Bray, A.M., and Geysen, M.H. (1990). Multi-pin peptide synthesis strategy for T cell determinant analysis. J. Immunol. Meth. *134*, 23–33.
Maeji, N.J., Tribbick, G., Bray, A.M., and Geysen, M.H. (1992). Simultaneous multiple synthesis of peptide-carrier conjugates. J. Immunol. Meth. *146*, 83–90.
Meldal, M., Holm, C.B., Bojesen, G., Jakobsen, M.H., and Holm, A. (1993). Multiple column peptide synthesis. Int. J. Peptide Protein Res. *41*, 250–260.
Meldal, M., Svendsen, I., Breddam, K., and Auzanneau France-I. (1994). Portion-mixing peptide libraries of quenched fluorogenic substrates for complete subsite mapping of endoprotease specificity. Proc. Natl. Acad. Sci. USA *91*, 3314–3318.
Moos, W.H., Green, G.D., and Pavia, M.R. (1993). Recent advances in the generation of molecular diversity. In: Annual Report Medicinal Chemistry, (Academic Press: San Diego, CA), pp. 315–324.
Needels, M.C., Jones, D.G., Tate, E.H., Heinkel, G.L., Kochersperger, L.M., Dower, W.J., Barrett, R.W., and Gallop, M.A. (1993). Generation and screening of an oligonucleotide-encoded synthetic peptide library. Proc. Natl. Acad. Sci. USA *90*, 10700–10704.
Nielsen, J., Brenner, R.A., and Janda, K.D. (1993). Synthetic methods for the implementation of encoded combinatorial chemistry. J. Am. Chem. Soc. *115*, 9812–9813.
Nikolaiev, V., Stierandova, A., Krchňák, V., Seligman, B., Lam, K.S., Salmon, S.E., and Lebl, M. (1993). Peptide-encoding for structure determination of nonsequenceable polymers within libraries synthesized and tested on solid-phase supports. Pept. Res. *6*, 161–170.
O'Neil, K.T., DeGrado, W.F., and Hoess, R.H. (1994). Phage display of random peptides on a protein scaffold. In: Techniques in Protein Chemistry, J.W. Crabb, ed. (Academic Press: San Diego), pp. 517–524.
Ohlmeyer, M.H., Swanson, R.N., Dillard, L.W., Reader, J.C., Asouline, G., Kobayashi, R., Wigler, M., and Still, W.C. (1993). Complex synthetic chemical libraries indexed with molecular tags. Proc. Natl. Acad. Sci. USA *90*, 10922–10926.
Parmley, S.F., and Smith, G.P. (1989). Filamentous fusion phage cloning vectors for the study of epitopes and design of vaccines. Adv. Exp. Med. Biol. *251*, 215–218.
Patek, M., Drake, B., and Lebl, M. (1994). All-cis cyclopentane scaffolding for combinatorial solid phase synthesis of small non-peptide compounds. Tetrahedron Lett. *35*, 9169–9172.
Pavia, M.R., Sawyer, T.K., and Moos, W.H. (1993). The Generation of Molecular Diversity. Bioorg. Med. Chem. Lett. *3*, 387–396.Ramachandran, G.N., and Sassisekharan, V. (1968). Conformation of polypeptides and proteins. Adv. Prot. Chem. *28*, 283–437.
Rinnova, M., Jezek, J., Malon, P., and Lebl, M. (1993). Comparative Multiple Synthesis of 50 Linear Peptides -Evaluation of Cotton Carrier vs T-Bag-Benzhydrylamine Resin. Pept. Res. *6*, 88–94.
Salmon, S.E., Lam, K.S., Lebl, M., Kandola, A., Khattri, P.S., Wade, S., Patek, M., Kocis, P., Krchňák, V., Thorpe, D., and Felder, S. (1993). Discovery of biologically active peptides in random libraries: Solution-phase testing after staged orthogonal release from resin beads. Proc. Natl. Acad. Sci. USA *90*, 11708–11712.
Scott, J.K., and Smith, G.P. (1990). Searching for peptide ligands with an epitope library. Science *249*, 386–390.
Scott, J.K. (1992). Discovering peptide ligands using epitope libraries. TIBS *17*, 241–245.
Scott, J.K., and Craig, L. (1994). Random peptide libraries. Current Opinion Biotechnol. *5*, 40–48.
Sebestyen, F., Dibo, G., Kovacs, A., and Furka, A. (1993). Chemical synthesis of peptide libraries. Bioorg. Med. Chem. Lett. *3*, 413–418.
Sepetov, N.F., Issakova, O.L., Lebl, M., Swiderek, K., Stahl, D.C., and Lee, T.D. (1993). The use of hydrogen-deuterium exchange to facilitate peptide sequencing by electrospray tandem mass spectrometry. Rapid Commun. Mass Spectrom. *7*, 58–62.
Simon, R.J., Kaina, R.S., Zuckermann, R.N., Huebner, V.D., Jewell, D.A., Banville, S., Simon, N.G., Wang, L., Rosenberg, S., Marlowe, C.K., Spellmeyer, D.C., Tan, R., Frankel, A.D., Santi, D.V., Cohen, F.E., and Bartlett, P.A. (1992). Peptoids: A modular

approach to drug discovery. Proc. Natl. Acad. Sci. USA *89,* 9367–9371.
Simon, R.J., Martin, E.J., Miller, S.M., Zuckermann, R.N., Blaney, J.M., and Moos, W.H. (1994). Using peptoid libraries [oligo N-substituted glycines] for drug discovery. In: Techniques in Protein Chemistry. J.W. Crabb, ed. (Academic Press: San Diego), pp. 533–539.
Stankova, M., Issakova, O., Sepetov, N.F., Krchňák, V., Lam, K.S., and Lebl, M. (1994). Application of one-bead-one-structure approach to identification of nonpeptidic ligands. Drug Development Research, *33,* 146–156.
Vagner, J., Krchňák, V., Sepetov, N.F., Strop, P., Lam, K.S., Baran", G., and Lebl, M. (1994a). Novel methodology for differentiation of "surface" and "Interior" areas of poly(oxyethylene)-polystyrene (PEE-PS) supports. In: Innovation and Perspectives in Solid Phase Synthesis. R. Epton, ed., pp. 347–352.
Vagner, J., Krchňák, V., Sepetov, N.F., Strop, P., Lam, K.S., Barany, G., and Lebl, M. (1994b). Enzyme-mediated spatial segregation on individual polymeric support beads, and application to generation of encoded combinatorial libraries., in preparation.
Zuckermann, R.N., Kerr, J.M., Kent, S.B.H., and Moos, W.H. (1992). Efficient method for the preparation of peptoids [oligo(N-substituted glycines)] by submonomer solid-phase synthesis. J. Am. Chem. Soc. *114,* 10646–10647.

3 Peptide Libraries Bound to Continuous Cellulose Membranes: Tools to Study Molecular Recognition

Jens Schneider-Mergener, Achim Kramer and Ulrich Reineke

3.1 Introduction

There are different strategies for the construction of soluble and solid phase-bound peptide libraries. These libraries have been used for the detection of antibody epitopes as well as peptide antagonists of medically relevant proteins (for review see Jung and Beck-Sickinger, 1992; Houghten, 1993; Pavia et al., 1993; Gallop et al., 1994 and literature cited therein). We have prepared structurally different combinatorial peptide libraries by spot synthesis (Frank, 1992) which is a rapid and simple tool for the simultaneous and positionally addressable synthesis of thousands of peptides or peptide mixtures bound to continuous cellulose membrane supports. These libraries were used for the detection of peptides that bind to proteins, metals and nucleic acids (Kramer et al., 1993; Kramer et al., 1994; Kramer et al., 1995; Volkmer-Engert et al., 1994; Malin et al., 1995; Winkler et al., 1995; Kramer and Schneider-Mergener, 1995; Stigler et al., 1995). The identified peptides could serve as lead compounds for therapeutical compounds or for diagnostic purposes.

Here we report a further optimization of cellulose-bound peptide libraries. As a model system of peptide-protein interaction we studied the molecular base of antibody-peptide recognition by synthesizing and screening different types of combinatorial peptide libraries. In addition, mutational analyses of linear peptide epitopes, substituting each single epitope residue by all 20 gene-encoded amino acids, gave interesting insights into the molecular nature of peptide-antibody recognition. The information gained, resulted in the synthesis of a combinatorial peptide library that is possibly able to detect any linear peptide epitope recognized by its antibody. These libraries might then also be used for the detection of peptides that specifically bind to any given protein molecule. Furthermore, the use of cellulose-bound peptide libraries for the identification of metal binding peptides is demonstrated here for the detection of lead(II) binding peptides.

3.2 Detection of antibody epitopes

We started with the synthesis of a combinatorial hexapeptide library XXB_1B_2XX (Geysen et al., 1986, Houghten et al., 1991). At the X-position an amino acid mixture consisting of 17 amino acids (cysteine, methionine and tryptophane were omitted) was simultaneously incorporated, whereas at the B-positions one of the twenty amino acids was coupled. Thus, the complete library contains 400 (20^2) peptide mixtures (spots) and consisits of roughly 33.4 million (400 x 17^4) single peptides. After incubation with the monoclonal antibody Tab2 (Hoeprich et al., 1989), we were able to detect and characterize its peptide epitope located within the amino terminal region NH_2-VVSHFNDCP- of transforming growth factor alpha (TGFα) (Kramer et al., 1993, 1994). However, several other epitopes located within in the preS2 region of hepatitis B virus (Kramer and Schneider-Mergener, 1995), the capsid protein p24 of HIV-1 (Höhne et al., 1993) and the transmembrane protein gp41 of HIV-1 (Porstmann et al., 1990) could not be detected with this type of XXB_1B_2XX library. These epitopes consist of 9, 11 and 13 residues, respectively.

Recent automation (Abimed, Langenfeld, Germany) of spot synthesis, enabled us to synthesize a library $XXXXB_1B_2B_3XXXX$ consisting of 8000 (20^3) peptide mixtures (spots) (Kramer and Schneider-Mergener, 1995). This library theoretically contains 5.6×10^{13} ($20^3 \times 17^8$) single peptides. Despite the extremly complex nature of this library, a highly specific interaction of this library with Tab2 was observed (Figure 3.1). The three most intensive spots contain the FND, HFN and the SHF sequences found in the N-terminal region of TGFα indicating that the sequence SHFND is at least part of the TGFα epitope. From the sequences observed in the lower right part of this library, it also became evident that the phenylalanine of the epitope can be substituted by a tyrosine residue. The results obtained from the library are consistent with the data obtained from a mutational analysis of the TGFα-derived peptide VVSHFND (chapter 3, Figure 3.2) and antibody binding studies of epitope-derived sequences. Using a competitive ELISA with solid phase-adsorbed TFGα, we determined the dissociation constants of the peptides HFND ($K_D = 1 \times 10^{-5}$M), SHFND ($K_D = 3 \times 10^{-7}$M), VSHFND ($K_D = 8 \times 10^{-8}$M) and VVSHFND ($K_D = 9 \times 10^{-9}$M). Although the peptide VVSHFND was identified as being the best binding peptide, the two N-terminal valine residues can be substituted by all other residues without loss of binding (Figure 3.2), whereas deletion of aspartic acid completely abolishes binding.

We then incubated the library $XXXXB_1B_2B_3XXXX$ with the monoclonal antibody CB4–1 recognizing the epitope GATPQDLNTML located within the capsid

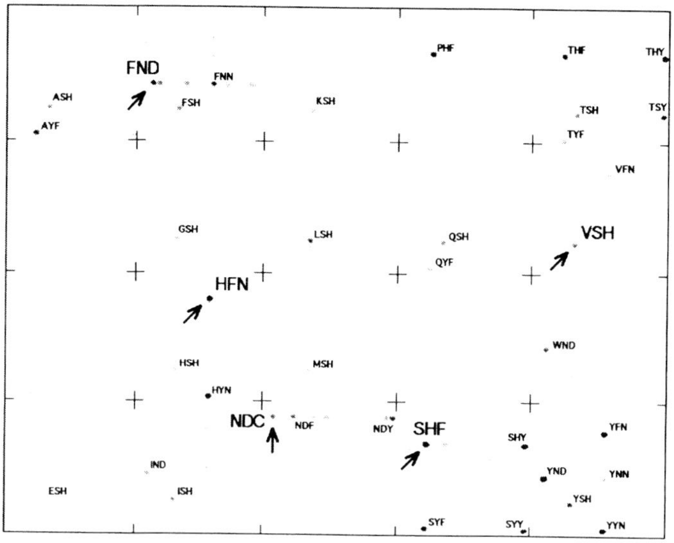

Figure 3.1: Binding of the antibody Tab2 to the library $XXXXB_1B_2B_3XXXX$
The antibody Tab2 recognizing the N-terminal region VVSHFNDC- of TGFα was incubated with a libary $XXXXB_1B_2B_3XXXX$. Binding was visualized as described using a second alkaline phosphatase-conjugated antibody (Kramer et al., 1994). The library was prepared by automated spot synthesis (Abimed, Langenfeld, Germany). For generation of the sequences the software program package DIGEN (Jerini Bio Tools GmbH, Berlin, Germany) was used. The library consists of 20^3 (8000) peptide mixtures (spots). The crosses divide the library into 20 squares, each containing 20^2 (400) spots. In each square B_2 and B_3 are defined as a 20 x 20 matrix (B_2 = rows, B_3 = columns), whereas B_1 is identical (one of the 20 amino acids starting with alanine (A) in the upper left and ending with tyrosine (Y) in the lower right square). TGFα-derived sequences are marked with arrows.

protein p24 of HIV-1 (Höhne et al., 1993). No clear cut result was obtained. Extremely weak binding to the spot with $B_1B_2B_3$ being DLN could be observed (see Figure 3.3). The low intensity of this spot was similar in comparison to several other spots most likely caused by unspecific binding and only marginally above background (data not shown). To investigate whether one would be able to detect this 11 amino acid residue epitope using a combinatorial library $XXXXB_1B_2B_3B_4XXXX$, we prepared eight peptide mixtures XXXXGATPXXXX, XXXXATPQXXXX, XXXXTPQDXXXX etc. comprising the complete epitope. Only the mixture XXXXDLNTXXXX was clearly recognized by CB4–1 (Kramer, unpublished results). However, due to experimental limitations, a complete library $XXXXB_1B_2B_3B_4XXXX$ consisting of 160,000 peptide mixtures (spots) would be an enormous synthetic effort at the moment. For that reason, we had to find a different approach for the detection of longer epitopes, rather than increase the number of defined positions.

3.3 Mutational analyses of peptide epitopes

To obtain more insight into the molecular nature of the peptide-antibody interaction, we performed mutational analyses of several antibody epitopes. This is demonstrated for the shorter TGFα epitope and the longer p24 epitope (Figure 3.2). We substituted each single position of both epitopes by all 20 amino acids. This approach allowed the identification of key residues in binding as well as

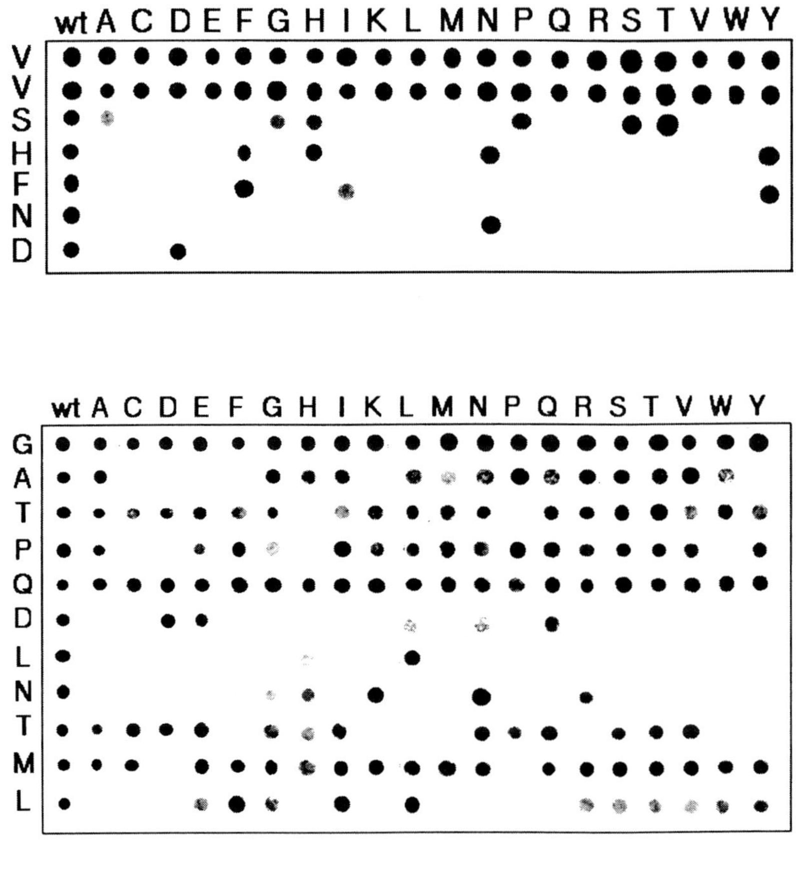

Figure 3.2: Mutational analyses of a TGFα and p24 epitope recognized by the antibodies Tab2 and CB 4–1
Each residue of the TGFα epitope VVSHFND recognized by Tab2 and the HIV-1 p24 epitope GATPQDLNTML bound by CB 4–1 is substituted by all 20 L-amino acids (rows) and analyzed for binding (see Figure 1, Volkmer-Engert et al., 1994). Each spot represents a single substitution analogue. All spots in the left column are identical and represent the wild-type (wt) sequence.

their location within the epitope. In the case of the TGFα epitope VVSHFND, two highly specific key residues, asparagine and aspartic acid at positions 6 and 7 were detected. Phenylalanine at position 5 could be substituted by the similar tyrosine and histidine at position 4 by phenylalanine, asparagine and tyrosine. Serine at position 3 could be replaced by the similar threonine and other small residues such as glycine, alanine and proline, but also histidine. Both valine residues could be substituted by all other residues without any loss of binding. Since the deletion of both valine residues significantly reduces the affinity of the peptide (see chapter 2) to its antibody Tab2, a backbone interaction of the first two residues to Tab2 seems to play an important role in binding. In the p24 epitope GATPQDLNTML, leucine at position 7 is most sensitive to substitution, followed by aspartic acid, asparagine, leucine and alanine at positions 6, 8, 11 and 2, respectively (Figure 3.2). The spot intensity on the membranes reflects the binding affinity of the corresponding soluble peptides (Volkmer-Engert et al., 1994).

From the mutational analyses of the epitopes shown here, as well as of several others, one clear result was obtained. The number of key residues in antibody epitopes generally varies between two and four. For instance, in the gp41 epitope CSGKLICTTAVPW recognized by the antibody 3D6 (Porstmann et al., 1990), isoleucine 6, threonine 8 and proline 12 were identified as key residues (Stigler et al., 1995). In the hepatitis B virus preS2 epitope QDPRVRGLY detected by the antibody 1F6 (Porstmann, unpublished results), glycine 7 and tyrosine 9 are key residues. In contrast to the p24 epitope, the preS2 epitope could be identified using a cellulose-bound combinatorial library XXXX$B_1B_2B_3$XXXX (Kramer and Schneider-Mergener, 1995). Another important feature of antibody-peptide recognition is the observation that the key residues can be distributed over the complete epitope, but often tend to cluster within a few residues (four to six). This was also observed by other groups using different approaches to characterize antibody epitopes at the molecular level (Geysen et al., 1984; Houghten, 1985; Geysen et al., 1987; Appel et al., 1990; Pinilla et al., 1994).

3.4 Positional scanning combinatorial library

The information obtained from the mutational analyses described above concerning the number and location of key residues in peptide-antibody recognition was used to construct a positional scanning library (Pinilla et al., 1992) with three defined amino acids. In this library XXXX(3B3X)XXXX an internal core consisting of three defined and three random positions (in parentheses) contains all possible distance patterns of defined amino acids, for instance XXXX(X$B_1B_2B_3$XX)XXXX, XXXX(XB_1B_2XB_3X)XXXX, XXXX (XB_1B_2XXB_3)XXXX etc. Such a library theoretically consists of 80,000 (10 x 20^3) peptide mixtures, if one distance pattern of the

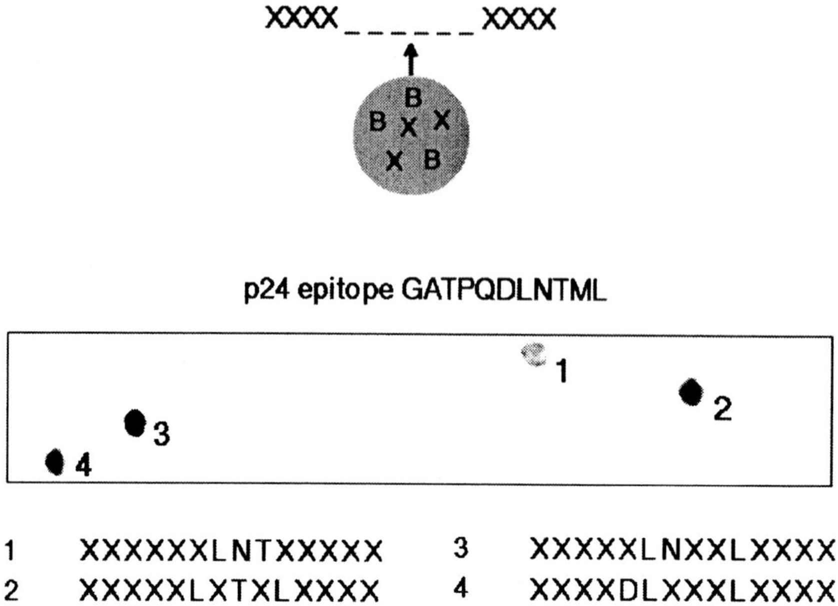

Figure 3.3: Detection of a p24 epitope
A positional scanning library XXXX(3B3X)XXXX was incubated with the antibody CB 4–1 recognizing the p24 epitope GATPQDLNTML and binding visualized as described in Figure 3.1. The internal hexapeptide core (in parentheses) contains all possible distance patterns for the three defined amino acids (see text). Only epitope-derived patterns were synthesized, resulting in 70 possible peptide mixtures (spots). The sequences of CB 4–1 binding spots are given.

defined amino acids only occurs once within the hexapeptide core sequence. For instance XXXX(X$B_1B_2B_3$XX)XXXX is allowed, but not the other possible sequences XXXX($B_1B_2B_3$XXX)XXXX, XXXX(XX$B_1B_2B_3$X)XXXX and XXXX(XXX$B_1B_2B_3$)XXXX.

In order to test whether such a library would allow the detection of the p24 epitope using the antibody CB 4–1, we synthesized this library with only p24 epitope-derived sequences at the B-positions (Figure 3.3). In contrast to the XXX$B_1B_2B_3$XXXX. library, this library now clearly allowed the detection of this 11 amino acid epitope. The most intensive signal, spot 3, corresponds to the sequence XXXXXLNXXLXXXX. The other spots also contain leucine 7 of the epitope. Here again, the results obtained from the positional scanning library correlate well with those from the mutational analysis of the p24 epitope in which leucine 7 appeared to be the key residue most sensitive to substitution (Figure 3.2).

Having obtained these results, we then synthesized the complete library XXXX(3B3X)XXXX consisting of 68,000 peptide mixtures (cysteine omitted). After incubation of the library with the antibody CB 4–1 one surprising result was observed. Besides the detection of p24-derived sequences also other binding of motifs occurred (Figure 3.4). Spots 1–5 can be attributed to the wild-type motif GATPQDLNTML, whereas spots 6–8 represent a novel motif which is bound by CB 4–1. The sequences of spots 1–5 can be explained by the mutational analysis the p24 epitope (Figure 3.2) in which leucine 7, asparagine 8 and leucine 11 are key residues in binding (in the order of importance). An algorithm for the determination of the relative order of importance of each epitope residue was developed (Volkmer-Engert et al., 1994). In spots 1–5 at least one of these key residues is present. In spot 1 and 2 threonine 9 is replaced by glutamine or asparagine, in spot 3 methionine 10 is substituted by lysine etc. indicating that these substitu-

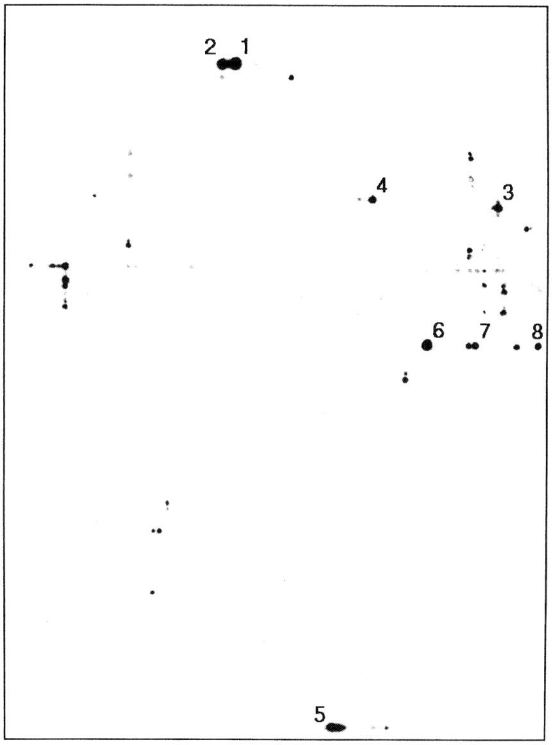

Figure 3.4: Positional scanning combinatorial library XXXX(3B3X)XXXX
Binding of the antibody CB 4–1 to the positional scanning library XXXX(3B3X)XXXX (68,000 spots in total) is demonstrated. From the complete library only a part (10,800 spots) is shown. The sequences of spots 1–8 are: 1 XXXX(XNQXLX)XXXX, 2 XXXX(XNNXLX)XXXX, 3 XXXX(XLNXKX)XXXX, 4 XXXX(XKQXLX)XXXX, 5 XXXX(XNXKLX)XXXX, 6 XXXX(XGAXIX)XXXX, 7 XXXX(XGKXIX)XXXX, 8 XXXX(XGVXIX)XXXX. In contrast to Figures 1–3 a chemoluminescence detection system (Amersham, Braunschweig, Germany) was applied using a peroxidase-conjugated second antibody.

tions increase binding affinities. The results obtained again clearly reflect those observed in the mutational analysis experiments of the p24 epitope. The remaining X positions of spots 1–5 were then further defined by synthesizing additional libraries. Using this approach we identified a peptide **GATPEDLNQKLAGNK** (wild-type epitope sequences in bold letters) that binds CB 4–1 with higher affinity ($K_D = 9 \times 10^{-10}$M) than the wild-type epitope GATPQDLNTML ($K_D = 4 \times 10^{-9}$M). Spots 6–8 represent a second motif GLYEWGGARITNTD in which tryptophan 5, glycine 6 and 7 as well as isoleucine 10 are key residues. This motif binds CB 4–1 with similar affinity (8×10^{-8} M) compared to the wild-type epitope. Since these two peptides compete for binding CB 4–1, they must bind to an at least similar site on the hypervariable surface of this antibody. Additional binding motifs could identified pointing to the molecular origin of cross reactivity of monoclonal antibodies (a detailed analysis of these results will be published elsewhere).

3.5 Identification of metal binding peptides

As already described, we used combinatorial cellulose-bound libraries for the detection of silver, technetium and nickel binding peptides (Kramer et al., 1993; Kramer et al., 1994; Malin et al., 1995). Figure 3.5 displays the binding patterns for six metals, such as technetium, calcium, iron, nickel, lead and silver each incubated with the library XB_1XB_2XX. The binding patterns for all metals clearly differ. Technetium has a strong preference for cysteine containing peptides, whereas nickel prefers lysine, cysteine and histidine, but also arginine. Calcium prefereably binds aspartic and glutamic acid containing peptide mixtures, as expected. Iron prefers histidine, lysine, and arginine but does not bind cysteine containing peptide mixtures under the conditions applied. Silver clearly prefers histidine, but also cysteine and methionine. In the case of lead, cysteine, aspartatic and glutamic acid peptide mixtures are equally bound. The difference in the binding patterns for each metal indicates that one could be able to identify peptides or derivatives that not only bind metals, but also could distinguish between them. This might lead to potential medical applications as well as a possible use in environmental technology.

Here we report the identification and characterization of a lead(II) (Pb^{2+}) binding peptide. In an initial screen, we incubated the library XB_1XB_2XX with lead acetate (Figure 3.5). There are several possibilities for analyzing the contribution to binding of single amino acids of the randomized positions. One possibility would be the selection of a single spot and subsequent iterative definition of the remaining X-positions (Geysen et al., 1986; Houghten et al., 1991; Kramer et al., 1994). Based on the binding pattern observed for the library incubated with lead,

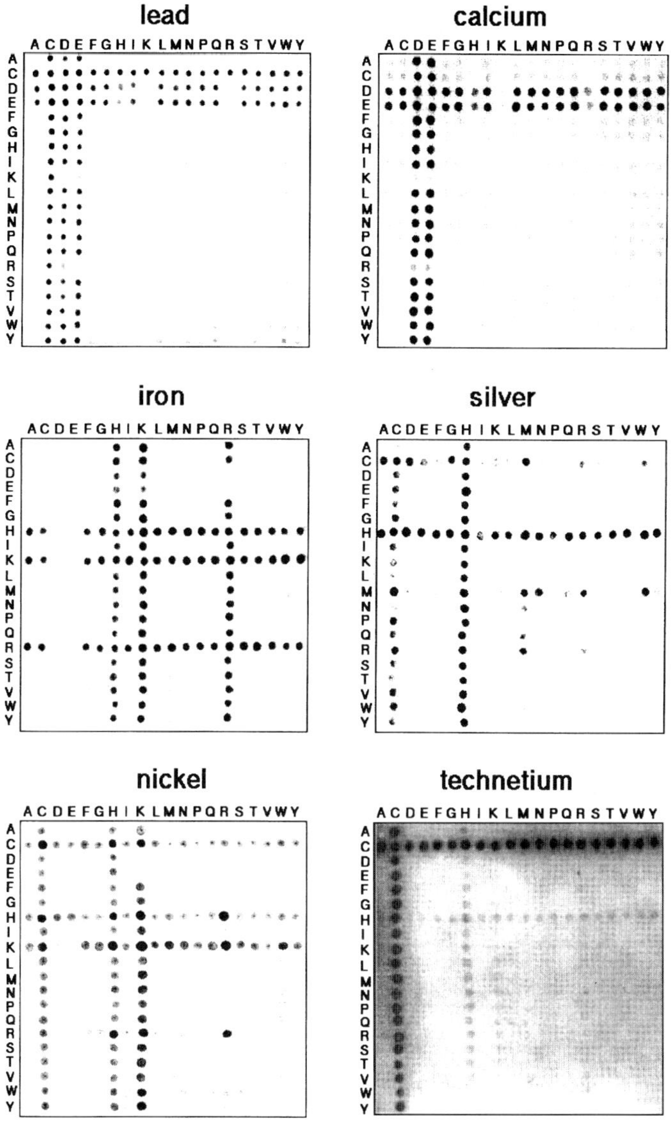

Figure 3.5: Binding of different metals to XB_1XB_2XX libraries
Incubation of the libraries with silver(I), technetium-99m and nickel(II) was performed as described (Kramer et al., 1993, 1994). In the case of iron(III) and calcium(II) the two isotopes ^{55}Fe (81 mCi, 1–10 µCi/mg) and ^{45}Ca (0.3 µCi $CaCl_2$, 3.98 mCi/mg) (Amersham, Braunschweig, Germany) were applied and the membranes subsequently analyzed using a PhosphoImager (Molecular Dynamics, Sunnyvale, USA). To identify lead binding peptides, the library was incubated with a 2 mM solution of $Pb(CH_3COO)_2$ in 20 mM Tris buffer (pH 7.0) for 30 min. Binding of lead(II) was visualized by precipitating lead with H_2S (B_1 = rows, B_2 = columns).

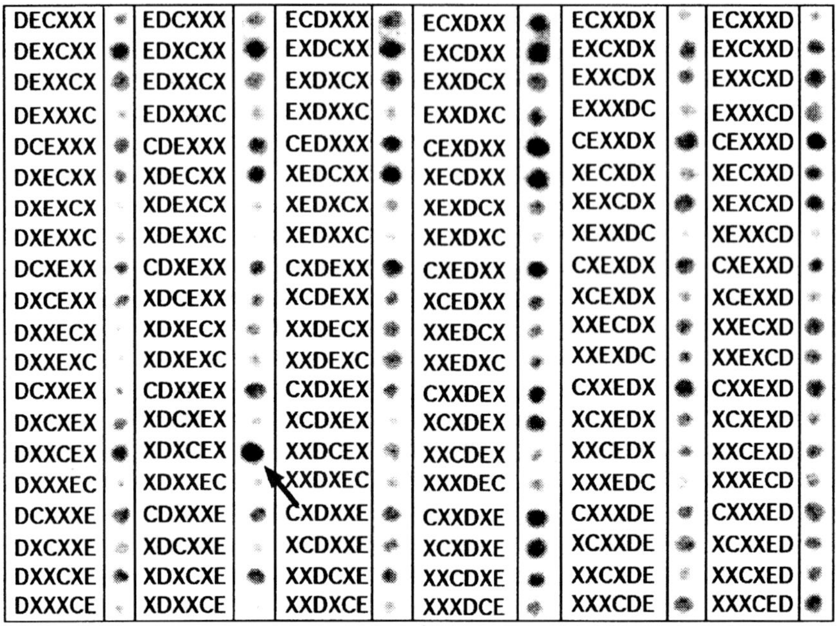

Figure 3.6: Identification lead binding peptide mixtures using a positional scanning library. Each spot represents a hexapeptide mixture containing one cysteine, one asparatic and one glutamic acid as well as three randomized positions (X). In the X-positions cysteine, aspartic and glutamic acid were omitted as described in chapter 5. All combinations possible (120) were synthesized and subsequently analyzed for lead binding as described in Figure 3.5. The peptide mixture XDXCEX was selected for the iterative identification of single lead binding hexapeptides (Figure 3.7).

we decided to apply an alternantive strategy. We started with a positional scanning library synthesizing all possible combinations of one cysteine, one glutamic and one aspartic acid residue within a hexapeptide sequence containing three randomized positions. The complete library consisits of 120 peptide mixtures (Figure 3.6). The peptide mixture XDXCEX appeared to be the best binding spot. In the second screening step, the two C-terminal randomized positions were defined by synthesizing a library XDB_1CEB_2 (Figure 3.7). Since the difference in lead binding to distinct spots could be either caused by a difference in affinity or stoicheiometry of binding, we excluded cysteine, aspartate and glutamate at the B positions in the second screen in order to avoid the creation of a possible second lead binding site within the same peptide. The other possibility that one metal ion is bound by more than one peptide can be excluded in this case due to the fixation and separation of the peptides on the membrane.

After selection of 21 peptide mixtures from the library XDB_1CEB_2, all possible 357 single peptides were synthesized on cellulose to study the contribution of the

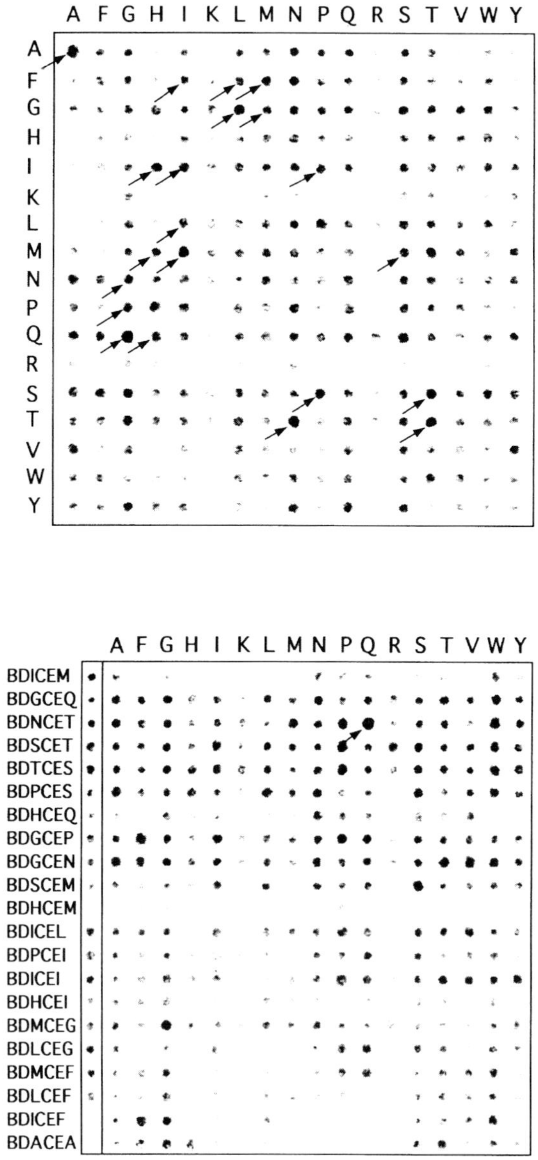

Figure 3.7: Identification of lead(II) binding peptides
After synthesis and incubation of the library XDB_1CEB_2 with lead, 21 of the best binding peptide mixtures (marked with arrows in the upper library) were selected for defining the remaining X-position (B_1 = columns, B_2 = rows; cysteine, aspartatic and glutamic acid were omitted). In the left column of the lower library, the 21 starting peptide mixtures are shown. Each spot of the library represents a single hexapeptide. The best binding peptide QDNCET was chosen for the characterization of lead binding in solution (Figures 3.8 and 3.9).

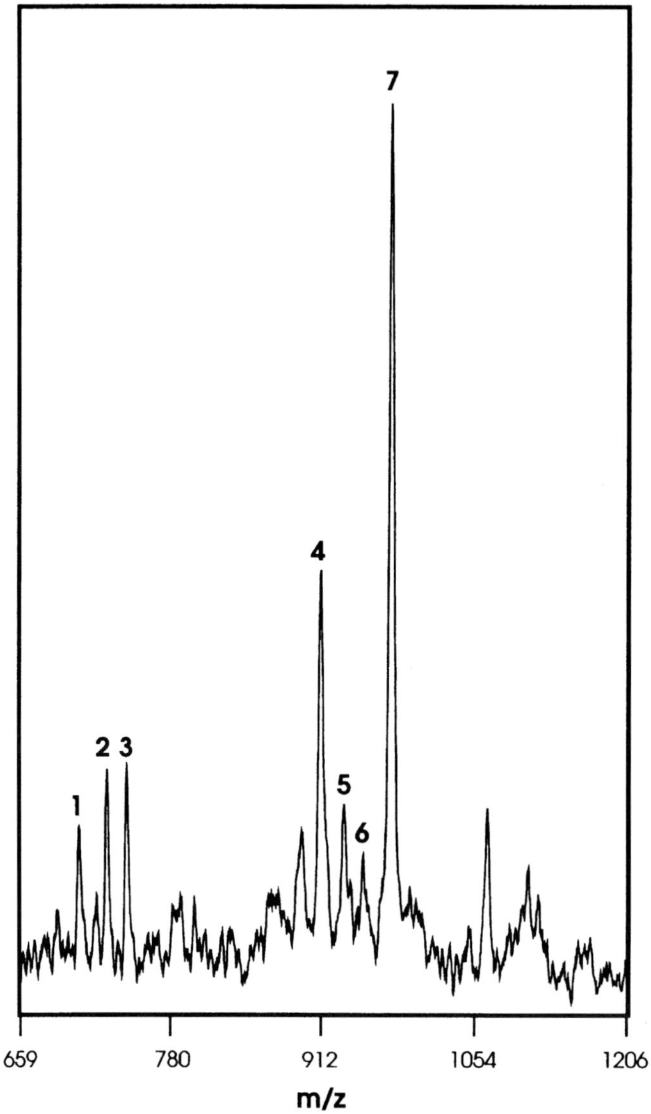

Figure 3.8: Laser desorption time-of-flight mass spectrometry of the peptide QDNCET in the presence of lead(II).
After incubation of the peptide QDNCET (1 mM) with $Pb(CH_3COO)_2$ (2 mM in H_2O) for 30 min, the mixture was analyzed by laser desorption time-of-flight mass spectrometry using α-cyano-4-hydroxy-cinnamic acid (LaserTec fOCUS, GSG/VESTEC, Karlsruhe, Germany). The free peptide (peaks 1–3) as well as four other peaks (4–7) representing complexes with one lead ion were observed. The observed ions are: 1 $[M+H]^+$, 2 $[M+Na]^+$, 3 $[M+K]^+$, 4 $[M+Pb]^+$, 5 $[M+Pb+Na]^+$,6 $[M+Pb+K]^+$, 7 $[M+Pb+Ca+Na]^+$.

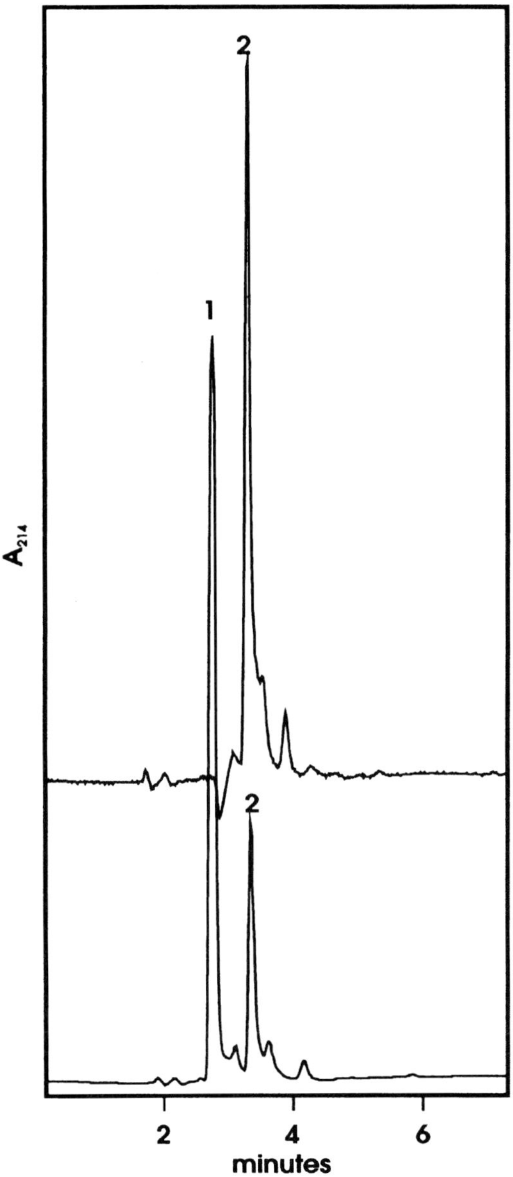

Figure 3.9: Detection of lead (II) binding to the peptide QDNCET by HPLC analysis. After incubation of the peptide QDNCET (1 mM) with $Pb(CH_3COO)_2$ (1 mM in H_2O) for 30 min, the mixture was analyzed by HPCL on a Vydac C18 column using a linear gradient 5% to 60% acetonitrile/water (0.05% TFA) for 20 min at 1.2 ml/min flow rate (lower panel). The peptide alone is shown in the upper panel. Peak 1 represents the peptide-lead complex, peak 2 the free peptide.

single amino acids in the randomized position (cysteine, aspartate and glutamate were omitted again) (Figure 3.7). The strongest lead binding peptide QDNCET was further analyzed for its lead binding properties in solution by laser desportion time-of-flight mass spectrometry (Figure 3.8) and HPLC analysis (Figure 3.9). The data obtained clearly demonstrate that this peptide binds only one lead ion.

3.6 Summary

We report the synthesis of different peptide libraries bound to continuous cellulose membrane supports. These libraries were used to study the peptide-antibody interaction as a model system for peptide-protein recognition. We were able to demonstrate that a positional scanning library XXXX(3B3X)XXXX enabled the detection of an 11 residue p24 epitope by the antibody CB 4–1 that could not be detected with a combinatorial library $XXXXB_1B_2B_3XXXX$. Other peptides that bound to the hypervarible regions of this antibody were identified demonstrating the molecular origin of cross reactivity of monoclonal antibodies. Experiments to use this type of positional scanning library for the detection of medically relevant protein molecules such as tumor necrosis factor alpha (TNFα) or transforming growth factor beta (TGFβ) are in progress. In contrast to the preparation of soluble peptide libraries, which is time consuming and expensive, automated spot synthesis allows the rapid generation of tens of thousands of peptides or peptide mixtures that can also be released from the membrane for screening purposes (Bray et al., 1991, 1992; Hoffmann and Frank, 1993). Besides proteins and metals, these libraries can be screened with a variety of other biologically interesting ligands such as nuleic acids or small organic molecules potentially to obtain peptides or derivatives that specifically bind these molecules, thus being of considerable interest in medicine, chemistry and biology. The difference in the binding behaviour of a given ligand to large numbers of peptide variants gives interesting insights into molecular recognition events and can be used to support computer docking experiments (Stigler et al., 1995). Furthermore, this approach offers new opportunities to study protein-protein interaction for instance involved in signal transduction (Heinrich et al., 1995) or heat shock protein-substrate recognition (Buckau and Schneider-Mergener, manuscript in prepration). Moreover, cellulose-bound peptide libraries are a powerful alternative to phage display peptide libraries (Kramer et al., 1995). Further optimization of these libraries with respect to reduction of combinatorial explosion (Kramer et al., 1995), structural diversification and fully automation of the synthesis process will be a challenging task for the near future.

Acknowledgements

We thank R. Geßner for assistance in the Ca and Fe binding experiments; J. Hellwig, B. Hoffmann and Ch. Landgraf for excellent technical assistance. This work was supported by the BMBF, DFG and the Fonds der Chemischen Industrie.

Literature

Appel, J.R., Pinilla, C., Niman, H., and Houghten, R.A. (1990). Elucidation of discontinuous linear determinants in peptides. J. Immunol. *144*, 975–983.

Bray, A.M., Maeji, N.J., and Geysen, H.M. (1990). The simultaneous multiple production of solution phase peptides; assessment of the Geysen method of simultaneous peptide synthesis. Tetrahedron Lett. *31*, 5811–5814.

Bray, A.M., Maeji, N.J., Jhingran, A.G., and Valerio, R.M. (1991). Gas phase cleavage of peptides from a solid support with ammonia vapour. Application in simultaneous multiple peptide synthesis. Tetrahedron Lett. *32*, 6163–6166.

Frank, R. (1992). Spot synthesis: an easy technique for the positionally addressable, parallel chemical synthesis on a membrane support. Tetrahedron *48*, 9217–9232.

Gallop, M.A., Barret, R.W., Dower, W.J., Fodor, S.P.A., and Gordon, E.M. (1994). Applications of combinatorial technologies to drug discovery. 1. Background and peptide combinatorial libraries. J. Med. Chem. *37*, 1233–1251.

Geysen, H.M., Meloen, R.H., and Barteling, S.J. (1984). Use of peptide synthesis to probe viral antigens for epitopes to a resolution of a single amino aicd. Proc. Natl. Acad. Sci. *81*, 3998–4002.Geysen, H.M., Rodda, S.J., and Mason, T.J. (1986). A priori delineation of a peptide which mimics a discontinuous antigenic determinant. Mol. Immunol. *23*, 709–715.

Geysen, H.M., Rodda, S.J., Mason, T.J., Tribbeck, G., and Schoofs, P.G. (1987). Strategies for epitope analysis using peptide synthesis. J. Immunol. Meth. *102*, 259–274.

Heinrich, P.C., Graeve, L., Rose-John, S., Schneider-Mergener, J., Dittrich, E., Erren, A., Herhartz, C., Hemmann, U., Lütticken, C., Wegenka, U.M., Weiergräber, O. and Friedemann, H. (1995). Membrane-bound and soluble interleukin-6 receptor: studies on structure, regulation of expression and signal transduction. Ann. New York Acad. Sci. in press.

Höhne, W.E., Küttner, G., Kießig, S., Hausdorf, G., Grunow, R., Winkler, K., Wessner, H., Gießmann, E., Stigler, R., Schneider-Mergener, J., v.Baehr, R. and Schomburg, D. (1993). Structural base of the interaction of a monoclonal antibody against p24 of HIV-1 with its peptide epitope. Mol. Immunol. *30*, 1213–1221.

Hoeprich Jr., P.D., Langton, B.C., Zhang, J.-W., and Tam, J.P. (1989). Identification of immunodominant regions of transforming growth factor α. J. Biol. Chem. *264*, 19086–19091.

Hoffmann, S., and Frank, R. (1993). New types of reversible peptide-resin linkages for the direct release of peptides and peptide amides into aqueous buffers. Poster 78, 13th Am. Peptide Symp., Alberta, Canada.

Houghten, R.A. (1985). General method for the rapid solid-phase synthesis of large numbers of peptides: specificity of antigen-antibody interaction at the level of indvidual amino acids. Proc. Natl Acad. Sci. *82*, 5131–5135.

Houghten, R.A., Pinilla, C., Blondelle, S.E., Appel, J.R., Dooley, C.T., and Cuervo, J.H. (1991). Generation and use of synthetic peptide combinatorial libraries for basic research and drug discovery. Nature *354*, 84–86.

Houghten, R.A. (1993). Peptide libraries: criteria and trends. Trends Gen. *9*, 235–239.

Jung, G. and Beck-Sickinger, A.G. (1992). Multiple peptide synthesis: methods and their applications. Angew. Chem. Int. Ed. Engl. *31*, 367-383.

Kramer, A., Volkmer-Engert, R., Malin, R., Reineke, U. and Schneider-Mergener, J. (1993). Simultaneous synthesis of peptide libraries on single resin and continuous membrane supports: identification of protein, metal and DNA binding peptide mixtures. Peptide Res. *6*, 314–319.

Kramer, A., Schuster, A., Reineke, U., Malin, R., Volkmer-Engert, R., Landgraf, C. and Schneider-Mergener, J. (1994). Combinatorial cellulose-bound peptide libraries: screening tool for the identification of peptides that bind ligands with predefined specificity. METHODS (Comp. Meth. Enzymol.) *6*, 388–395.

Kramer, A., Vakalopoulou, E., Schleuning, W.-D. and Schneider-Mergener, J. (1995). A genearal route to fingerprint analyses of peptide-antibody interactions using a clustered amino acid peptide library: comparison with a phage display library. Mol. Immunol. *32*, 459–465.

Kramer, A. and Schneider-Mergener, J. (1995). Highly complex combinatorial cellulose-bound peptide libraries for the detection of antibody epitopes. In: Peptides 1994 (ESCOM, Leiden), 475–476.

Malin, R., Reineke, U., Höhne, W., Semmler, W., Steinbrecher, R. and Schneider-Mergener, J. (1995). Identification of technetium-99m and nickel binding peptides using combinatorial cellulose-bound peptide libraries. In: Peptides 1994, (ESCOM, Leiden), 477–478.

Pinilla, C., Appel, J.R., Blanc, P., Houghten, R.A. (1992). Rapid identification of high affinity peptide ligands using positionally scanning synthetic peptide combinatorial libraries. BioTechniques *13*, 901–905.

Pinilla, C., Appel, J.R., and Houghten, R.A. (1994). Investigation of antigen-antibody interactions using a soluble, non support-bound synthetic decapeptide library composed of four trillion (4×10^{12}) sequences. Biochem. J. *301*, 847–853.

Porstmann, T., Döpel, H., Grunow, R., Henklein, P., and v.Baehr, R. (1990). Detection of immunodominant highly conserved regions within the transmambrane protein of the human immunodeficiency virus 1 by site-directed ELISAs with synthetic peptides and their application for screening purposes. Z. Klin. Chem. *45*, 341–346.

Stigler, R., Rüker, F., Katinger, D., Elliot, G., Höhne, W., Henklein, P., Ho, J.X., Kramer, A., Nugel, E., Porstmann, T. and Schneider-Mergener, J. (1995). Characterisation of the interaction between a Fab fragment against gp41 of HIV-1 and its peptide epitope using a peptide epitope library and molecular modelling. Protein Eng. *8*, 471–480.

Volkmer-Engert, R., Ehrhard, B., Höhne, W., Ehrhard, B., Kramer, A., Hellwig, J. and Schneider-Mergener, J. (1994). Preparation, analysis and antibody binding studies of simultaneously synthesized soluble and cellulose-bound HIV-1 p24 peptide epitope libraries. Lett. Peptide Sci. *1*, 243–253.

Winkler, D., Schuster, A., Hoffmann, B. and Schneider-Mergener, J. (1995). Synthesis of cyclic peptide libraries bound to continuous cellulose membrane supports. In: Peptides 1994 (ESCOM, Leiden), 485–486.

B Nucleic Acids Libraries

4 *In Vitro* Selection of Nucleic Acid Sequences that Bind Small Molecules

Jon R. Lorsch and Jack W. Szostak

4.1 Introduction

Nucleic acids can fold into three-dimnesional structures that can recognize small molecules with much the same avidity and selectivity as protein receptors and enzymes. The purpose of this chapter is to review progress in the selection of RNA receptors, or aptamers, for small ligands, and to highlight what is known about the structures of these molecules.

4.2 Natural RNA Receptors

Specific binding sites for small molecules are found in naturally occuring RNAs. The group I introns, for example, bind guanosine nucleotides with a K_d of approximately 0.5 mM. (Herschlag and Cech, 1990) At least part of the binding site has been localized, and the interaction involves the formation of a base triple between a G-C base pair on the RNA and the guanosine nucleotide. (Herschlag and Cech, 1990) The binding is specific, and the introns can discriminate between a large number of guanosine and nucleotide analogues (Michel et al., 1989; Bass and Cech, 1984). Arginine, which has a similar array of hydrogen bonding groups to guanosine, also binds to the ribozyme. (Bass and Cech, 1986; Yarus, 1989) Arginine-RNA interactions are common in nature. For instance, the TAR RNA of the Human Immunodeficiency Virus (HIV) binds to the arginine-rich Tat protein, and to a number of smaller arginine containing peptides derived from Tat, with K_ds in the low nM range.(Yarus and Majerfeld, 1992; Weeks et al., 1990)

The TAR RNA also binds L-arginine specifically, with a Kd of 4 mM. (Calnan et al., 1991) Thus, both the group I introns and TAR are able to fold into structures that specifically recognize arginine. A number of the aminoglycoside antibiotics have been shown to bind to both ribosomal RNA and to the group I introns at specific sites. (Tao and Frankel, 1992; Ahsen v. and Schroeder, 1991; Ahsen v. et al., 1991) Recently, it has been shown that neomycin B and related molecules can compete with the HIV Rev protein for binding to the Rev Response Element (RRE). (Moazed and Noller, 1987) In all three cases, the antibiotic appears to bind at a specific site in the corresponding RNA. $I.C.^{50}$ values for these molecules range from $<1\,\mu M$ to 5 mM; K_ds are rarely reported, but appear to be in the same range. Inhibition of group I intron self-splicing by streptomycin has been shown to be competitive with guanosine, leading Davies, von Ahsen and Schroeder to propose that the antibiotic binds in the guanosine binding site (Tao and Frankel, 1992). It cannot be ruled out, however, that the competitive nature of the inhibition results from two mutually exclusive, but separate, sites (i.e., negative allostery), especially since a mutation that weakens guanosine binding leads to tighter streptomycin binding. It has been proposed that the aminoglycoside antibiotics bind to RNA at a common structural motif, related to the group I introns' guanosine/ arginine binding site. (Ahsen v. and Schroeder, 1991; Zapp et al., 1993) The evidence for this proposal is that the active aminoglycoside antibiotics contain guanidinium groups that would confer the same hydrogen bonding capabilities as guanosine and arginine, coupled with the fact that the antibiotics appear to bind to specific sites (i.e. structures) within the RNAs. Furthermore, the Rev Response Element binds both Rev, an arginine rich protein, and aminoglycoside antibiotics. A universal arginine-binding RNA motif has also been proposed because a number of RNA-recognizing proteins and peptides contain arginine-rich domains (e.g. bacteriophage antiterminator N proteins, Tat, Rev). There is, however, no experimental evidence for such a common RNA structure, and there is considerable evidence against its existence. (Davies et al., 1993) For example, the Tat and Rev peptides appear to have very different structures, (Mattaj, 1993; Calnan et al., 1991; Tan et al., 1993) suggesting that they recognize their cognate RNAs in different ways. Arginine, being positively charged and having an array of groups capable of hydrogen bonding, is uniquely suited for use by RNA-recognizing proteins. It seems more likely that it is the versatility of arginine, rather than the existence of a single arginine-binding RNA motif, that is the reason behind the existence of arginine rich domains in RNA-binding proteins.

It has also been proposed that because the 16S subunit of the ribosome and the group I introns both bind the aminoglycoside antibiotics specifically, that these RNAs are closely related evolutionarily and that the aminoglycoside antibiotics coevolved with functional RNAs. (Zapp et al., 1993) However, there is no reason to suppose that aminoglycoside antibiotics, which are produced only by a small number of organisms today, existed that early in evolution. A more plausible explanation is that this class of compounds, which, like arginine, are positively charged and possess an array of hydrogen bond donors and acceptors, are able to interact with a variety of different RNA structures in a specific manner. In support of this latter proposal, Ellington, Famulok and coworkers have selected for RNAs that bind the aminoglycoside antibiotics kanamycin and lividomycin and neomycin.(Puglisi et al., 1992; Ellington, Personal Communication). Sequence analysis has revealed that there are many different primary and secondary structures capable of binding each of these compounds with high affinity. Thus, there is no reason to suppose that there will be only one type of aminoglycoside binding site found in naturally occuring RNAs.

4.3 *In Vitro* Selection

The dual nature of RNA, which is capable of folding up into functional tertiary structures and also acting as an information carrier, has made it easy to devise schemes for the isolation of rare functional sequences. The earliest demonstrations that RNAs with desired binding properties could be selected from libraries of random RNAs were performed by Tuerk and Gold and by Ellington and Szostak. (Famulok, Personal communication; Tuerk and Gold, 1990) The technique of *in vitro* selection, or SELEX (Systematic Evolution of Ligands by Exponential Enrichment) consists of two phases (Figure 4.1). A pool of random or partially random sequence RNA molecules is subjected to a selection for a physical property, for example binding to a ligand. The functional molecules are separated from non-functional ones via this selection step. The selected molecules are amplified by reverse transcription followed by PCR and transcription, yielding a new pool of RNAs enriched in active sequences. This process of selection and amplification is repeated until the pool is sufficiently enriched in molecules with the desired property. (Ellington and Szostak, 1990) Using this technique, Ellington and Szostak isolated, from a pool of 10^{13} different molecules, RNAs that are capable of binding to dye affinity columns. (Tuerk and Gold, 1990) They termed nucleic acids that had been selected for binding to small molecules "aptamers", from the Latin word aptus, meaning "to fit". A number of these selected RNAs discriminate between structurally similar dye matrices, demonstrating again that RNA is capable of highly refined molecular recognition. The K_ds for the dye matrix were determined for two dye aptamers, and were found to be in the mid to high μM range (~100–600 μM). Surprisingly, the number of solutions to this recognition problem is quite large: in their library, Ellington and Szostak found between 10^3 and 10^7 RNA sequences that could bind each of the seven different dye matrices used. This fact suggests that in sequence space as a whole, there are many more active molecules. Ellington and Szostak estimated the number of molecules in 100-mer sequence space (10^{60}) that can bind these types of ligands to be approximately 10^{50}.

Using the same techniques, Famulok and Szostak isolated, from a pool of $\sim 5 \times 10^{14}$ different molecules, 100–1000 RNAs that bind to D-tryptophan-derivatized agarose. (Jenison et al., 1994) The best clone binds to the resin with a K_d of ~20 μM. No binding of this RNA was detected to L-tryptophan agarose demonstrating that this RNA binds in a stereospecific fashion. The $\Delta\Delta G^0$ for binding D-tryptophan agarose relative to L-tryptophan agarose is estimated to be ^{3} 3.9 kcal mol^{-1}.

These two selections both yielded RNAs that bind to ligands on solid supports, but not to the same ligands free in solution (although Famulok has recently shown that the tryptophan aptamers do bind N-acetyl-tryptophan, but not tryptophan, in solution. (Ellington, Personal Communication) Presumably, the reason for this is that the bound RNAs were eluted from the matrix with a denaturing wash, and thus there was no selection against RNAs that interacted with the matrix and the linker as well as with the ligand. In order to get around this problem, Sassanfar and Szostak used affinity elution to isolate an RNA that binds to adenosine triphosphate in solution. (Famulok and Szostak, 1992) The structure of this molecule is shown in Figure 4.2a. It binds ATP highly specifically: the

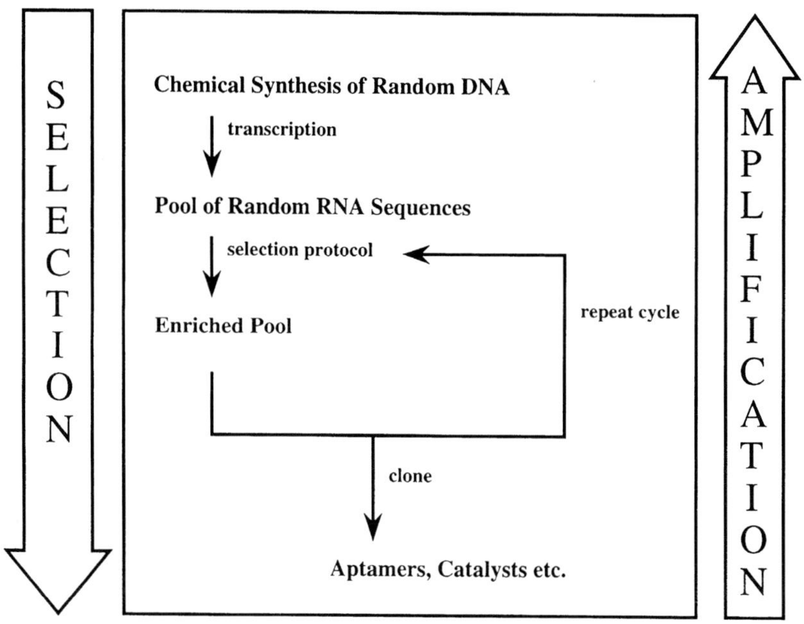

Figure 4.1: a) General outline of *in vitro* selection or SELEX. A library of random sequence nucleic acids is subjected to repeated cycles of selection for a desired function, followed by amplification to yield a pool of molecules enriched in the desired properties.

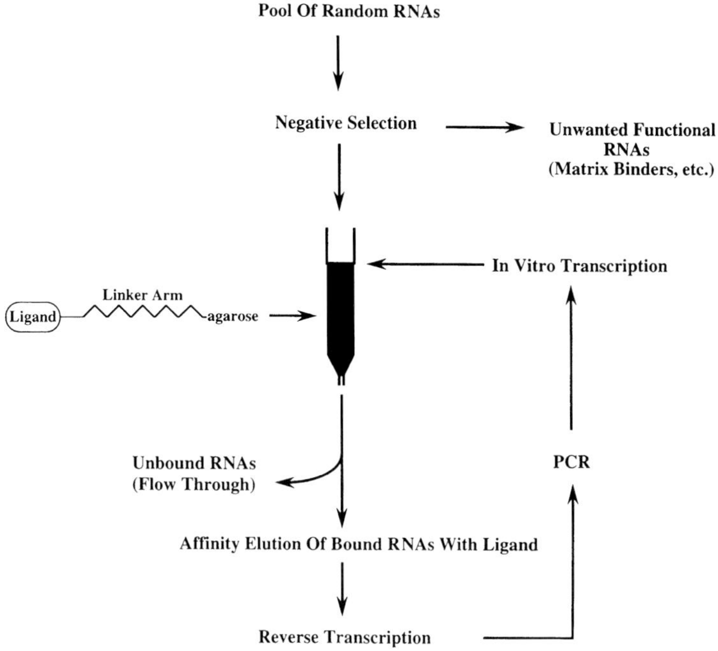

Figure 4.1: b) Outline of affinity chromatographic method of isolating RNA aptamers for small molecule ligands. A library of random sequence RNAs is subjected to a selection for binding to a matrix derivatized with a ligand. The matrix is then washed to remove non-specifically or weakly bound RNAs, and then molecules that bind the ligand in solution are eluted from the matrix by washing with a solution containing the ligand. In practice a negative selection is often performed prior to the affinity chromatography step, in order to remove molecules with undesirable properties (i.e., binding to the matrix and/or spacer arm, or binding compounds similar in structure to the ligand of interest). These negative selection steps can include precolumns or preelutions of bound molecules with other compounds that the aptamer should not bind (see, for example, Ellington and Szostak, 1990).

aptamer does not bind a variety of nucleotides and nucleosides (e.g., GTP, CTP, UTP, inosine, purine riboside, 2-aminopurine riboside), and methylation of adenosine at N1, C2, N6, or O3′ abolishes binding to the RNA. The aptamer binds suprisingly tightly, with a K_d of ~1–10 μM for ATP in solution.

4.4 Amino Acid Aptamers

The arginine molecule has proved of great interest to those involved in *in vitro* selection. One hope has been that a single arginine recognition motif would be elucidated by selecting for RNAs with the appropriate strength of interaction with the amino acid. Yarus and coworkers selected for aptamers that bind to arginine in solution with K_ds in the range of 0.5 mM. (Sassanfar and Szostak, 1993) They isolated three different motifs, none of which bears any resemblance to TAR RNA or to the guanosine/arginine binding site of the group I introns. The three motifs have somewhat different specificities for arginine analogues, suggesting that they may be structurally dissimilar. In other words, rather than having found a unique arginine binding motif that might begin to rationalize the genetic code, Yarus and coworkers have found that in a pool of 5×10^{13} different molecules there are at least three different RNA structures that can bind arginine with K_ds in the high micromolar to low milimolar range. In further experiments Connell and Yarus selected for an aptamer that binds guanosine, and a related aptamer that binds both arginine and guanosine. (Connell et al., 1993) Neither of these RNAs shows any obvious structural similarity to TAR, to the guanosine binding site of the group I introns, or to the arginine aptamers that they isolated in their earlier work (which, fortuitously, bind guanosine as well). Interestingly, the binding sites of this guanosine aptamer and the joint guanosine/arginine aptamer differed consistently at only one nucleotide, suggesting that this position dictates whether the molecule can bind only guanosine, or whether it binds both arginine and guanosine.

Famulok has also isolated arginine binding RNAs. He first selected for aptamers to L-citrulline, the urea analogue of arginine. (Connell and Yarus, 1994) He isolated a citrulline binding RNA motif, with a K_d for the amino acid in solution of~6μM. Based on this citrulline binding sequence he made a library of mutagenized RNAs, and selected from it RNAs that bind L-arginine. The new arginine aptamer binds arginine with a K_d of approximately 65μM, and is a triple mutant of the original L-citrulline receptor. Attempts to select for binders to L-albizziin, L-lysine, and L-glutamine from the library of mutagenized citrulline aptamers were unsuccessful, suggesting that solutions to these recognition problems were too far away in sequence space to be found in the library. As with the work of Yarus and coworkers, Famulok's arginine aptamer bears no resemblance to natural arginine recognition motifs, although it should be noted that his RNA binds significantly more tightly than do either TAR or the group I introns. It is clear from all of this work, however, that there are many different ways in which RNA can recognize molecules containing guanidinium and urea functionalities. There is no reason to think that Nature limited itself to just one solution. It is, of course, possible that while the primary and secondary structures of all of these arginine receptors are completely different, their tertiary structures are very similar. Based on how different the sequences and proposed secondary structures are, however, this seems unlikely.

The idea that RNA-amino acid interactions may have been involved directly in specifying the genetic code has sparked continuing interest in isolating RNAs that specifically recognize amino acids. Majerfeld and Yarus have recently reported the selection of an RNA with weak affinity for L-valine. (Famulok, 1994) The K_d for the affinity ligand, L-valinamide, is 3mM, and the K_d for L-valine is 12mM. The aptamer discriminates between L- and D-valine ($\triangle\triangle G^0$ = 1.7kcal/mol), and between L valine and other hydrophobic amino acids ($\triangle\triangle G^0$= 1.6–2kcal/mol). These sorts of interaction energies and discriminations, while intriguing, do not seem large enough to have led to an accurate genetic coding/decoding mechanism. Tighter and more specific aptamers for hydrophobic amino acids would give stronger support to the direct RNA-amino acid interaction model for the origin of the genetic code; in any case, these experiments argue against the significance of codon (or anticodon)-amino acid interactions, (Majerfeld and Yarus, 1994) since the sequences and structures needed to achieve even marginal recognition of amino acids are significantly larger than three nucleotides.

4.5 Cofactors

Selection of high-affinity aptamers for biological cofactors is of interest for several reasons. First, if the "RNA World" evolved to the point at which there were ribozymes that conducted many of the same catalytic tasks performed by today's protein enzymes, then RNA binding sites for enzyme cofactors would have been necessary to extend the limited chemical functionality available in RNA itself. (Yarus, 1993) Furthermore, RNA motifs that bind catalytically useful small molecules could be good starting points for selections for novel ribozymes (and this has, in fact, proven to be the case; see 34, 35). (Benner et al., 1990; Lorsch and Szostak, 1994) We have already mentioned the isolation of an aptamer for ATP, that binds its ligand with a K_d of 1μM. More recently, we have isolated an RNA that binds cyanocobalamin (vitamin B12) in solution with a K_d of 90nM, the lowest dissociation constant for a small molecule aptamer reported to date. (Wilson and Szostak, in press) This selection was carried out in 1MLiCl because the strong interaction of Li^+ with the phosphate backbone might allow the formation of otherwise inaccesible RNA tertiary structures. Indeed, binding is dependent on the presence of high concentraions of Li^+, and decreasing the Li+ concentration to 0.5M nearly abolishes binding. Using a library of mutagenized RNAs derived from the original aptamer sequence we were able to select for RNAs that bind cyanocobalamin in Na^+ or K^+ plus Mg^{2+}, but not in Li^+. It is not yet clear how (or whether) the two aptamers are structurally related to one another.

In another selection in our lab, using the same RNA library used to isolate the cyanocobalamin aptamer, Wilson and Szostak isolated a single RNA molecule (>90% of clones sequenced) that binds the cofactor biotin. (Lorsch and Szostak,

1994) The aptamer binds its ligand with a K_d of 10μM, and is highly specific. This RNA has been used as the starting point for a selection for self-alkylating ribozymes. (Lorsch and Szostak, 1994) Burgstaller and Famulok have recently reported the isolation of aptamers for the redox cofactors flavin mononucleotide (FMN) and flavin adenine dinucleotide (FAD). (Lorsch and Szostak, 1994) The FMN aptamer has a K_d for FMN of 0.5μM, and a K_d for FAD of 0.7μM. The FAD aptamer binds FMN with a K_d of 80μM, and to FAD with a K_d of 140μM. This RNA shows no affinity for adenosine, but binds flavin alone with a K_d of 20μM. Thus it appears that the aptamer recognizes just the isoalloxazine ring of the cofactors. Attempts were also made to select for aptamers for nicotinamide adenine dinucleotide (NAD) from the same random RNA library. All that was isolated, however, was the adenosine aptamer isolated previously by Sassanfar and Szostak. (Famulok and Szostak, 1992) This result represents the first time that the same RNA structural motif has been isolated from two different random RNA pools (and in two different laboratories). Interestingly, in the FAD selection, aptamers were isolated that were identical to one of the FMN binding RNA clones, again demonstrating the ability of these procedures to reproducibly isolate the same rare sequence from a pool of $>10^{14}$ sequences. Burgstaller and Famulok also tried to isolate nicotinamide mononucleotide (NMN) receptors, but were unsuccessful. (Lorsch and Szostak, 1994) This negative result may be a consequence of the fact that sequences that bind this ligand are so rare that they are only occasionally found in libraries with complexities of $\sim 10^{15}$; Lauhon and Szostak have successfully selected for an aptamer for NMN from a library of slightly greater complexity (Burgstaller and Famulok, 1994) This NMN aptamer binds both NAD and NMN, and has a K_d of approximately 2.5μM for NAD, but a K_d of nearly 40μM for NADH, showing that RNA can discriminate between redox states of the cofactor. They also isolated an aptamer for oxidized riboflavin that is structurally very different from the flavin aptamers isolated by Burgstaller and Famulok. (Lorsch and Szostak, 1994) Apparently different pools and/or different selection conditions (i.e., mono- and divalent cations used and their concentrations, pH, etc.) can yield very different results for the same molecular recognition problem. The selections described above demonstrate the versatility of RNA as a receptor for small molecule ligands. A recent report of RNA aptamers for the asthma drug theophylline demonstrates a potential application of such *in vitro* selections (Ellington and Szostak, 1990) RNAs were isolated that could discriminate by four orders of magnitude between theophylline and the closely related molecule caffeine (the two differ in that caffeine bears an extra methyl group). The aptamer binds theophylline with a K_d of 100 nM, but has a K_d of only 3.5mM for caffeine. This discrimination is almost certainly due to a steric conflict between the RNA and the N7 methyl group on caffeine. The levels of the drug theophylline must be measured carefully in patients, because at higher concentrations it can have serious toxic effects. Measurement of serum levels is complicated by the presence of caffeine and other related molecules. Thus, the theophylline aptamer, which discriminates between theophylline and caffeine 10-fold better than the antibody presently used for screening, could prove to have medical utility.

4.6 DNA Aptamers

After their initial work showing that it was possible to select for RNAs that specifically recognized dye affinity columns, (Tuerk and Gold, 1990) Ellington and Szostak went on to show that it was also possible to isolate single-stranded DNAs with similar functions. (Lauhon and Szostak, 1995) More recently, Huizenga and Szostak have selected for a DNA aptamer for adenosine and ATP. (Ellington and Szostak, 1992) The aptamer binds its ligands with a K_d of approximately 5–20 μM, and is highly specific. In neither case do the RNA versions of these receptors bind the ligand. It is expected that in the future more DNA aptamers for a wide variety of ligands will be isolated.

4.7 The Complexity of Complexity

How rare are functional nucleic acid molecules in sequence space? Table 4.1 gives some insight into to this question. In Ellington and Szostak's original work, (Tuerk and Gold, 1990) approximately 500 RNAs (out of a library of $\sim 10^{13}$ different molecules) were isolated that could bind the Cibacron Blue affinity resin, whereas $\sim 10^5$ sequences bound the Reactive Green 19 affinity resin. This suggests that there are a truly large number of molecules in sequence space that can bind dye affinity matrices. Interestingly, when they did the same selection using single-stranded DNA instead of RNA, they found that approximately the same fraction of sequences were functional, but the relative number of solutions to the Cibacron Blue vs. Reactive Green binding problems were reversed.

The dye affinity resins are ideally suited as ligands for nucleic acids: they contain large, planar aromatic surfaces against which nucleotide bases can stack (and intercalation may even be possible), and many hydrogen bond donors and acceptors. It is not too surprising, then, that there are a great many different ways that RNA and DNA can find to interact with them. As one would expect, as the ligands get smaller and the functionality less ideally suited to nucleic acid interaction, fewer solutions to the molecular recognition problems exist (Table 4.1).

The complexity, or informational content, of a binding motif should in theory be limited by the complexity of the library from which it was isolated. For example, a library of 10^{15} different molecules should contain nearly every possible 25-mer sequence ($4^{25} = 10^{15}$). Thus, the largest domain that one could expect to find in a pool of 10^{15} different molecules would have 25 defined bases. Even though it is hard to separate primary sequence requirements (i.e. absolutely conserved bases) from secondary ones (i.e. base pairs) no consensus reported to date is larger than would be expected based on the size of the libraries used. The chances of finding an active larger domain are quite small when one considers the size of sequence space: the number of possible 40-mers is $4^{40} \approx 10^{24}$ and thus a library of 10^{15} different molecules samples only 10^{-9} of 40-mer space (assuming the molecules are 40-mers or longer, that is). The chances of finding larger domains are obviously even smaller. How long should the RNA or DNA in a library be? The answer to this question is still not entirely clear. Longer molecules provide a linear increase in the probability of finding a defined sequence, and a combinatorial increase in the probability of finding two or more interacting sequences. Longer molecules could also fold into more complex structures than shorter ones. Recent work in our laboratory has suggested, however, that there are often "inhibitory sequences" attached to smaller functional domains in aptamers isolated from libraries of long RNAs. These inhibitory sequences presumably prevent proper folding of the RNA when not inactivated by other parts of the molecule. This inactivation usually involves base-pairing with non-essential sequences in the RNA.Thus, many active molecules may be lost from a large library because of poisoning by inhibitory sequences. It is conceivable that long pools, which can reach higher levels of complexity, may not yield as many solutions as shorter, less complex ones.

One final note regarding complexity issues. Examination of Table 1 will reveal that cumulative enrichment data over the course of a selection do not always do not always correlate well with pool complexity. Presumably this is due to hidden selective processes. For example, molecules that do not replicate well during the amplification step will tend to be under-represented, and it is even possible that many of the most structured molecules do not survive reverse-transcription and PCR. Molecules that are difficult to amplify, or which bind with less than 100% efficiency, will clearly require more enrichment than expected simply on the basis of their abundance in the original pool.

4.8 Aptamer Structures

A wide variety of secondary structures are used by the aptamers that have been isolated to date (Figure 4.2). Stem-loops, stem-bulge-stems, pseudoknots, and G-quartets are all represented in the handful of available aptamers. Little is known about the tertiary structures of these molecules, but presumably they too are

Table 4.1: Summary of Aptamer Selections

	Ligand	Library Complexity	# Unique Sequences	Consensus (Number)	Size of Domain[a]	K_d (μM)
RNA						
Ref. 24	Organic Dyes	~10^{13}	10^2–10^7	Some	22,28[b]	100–600
Ref. 26	Tryptophan Agarose	5 x 10^{14}	10^2–10^3	No	n.d.	20
Ref. 28	Arginine	5 x 10^{13}	10^4–10^5[c]	Yes (3)	15–19[d]	200–400
Ref. 27	ATP	~10^{14}	20–50	Yes (1)	13[e]	1
Ref. 35	Biotin	5 x 10^{14}	1	No	~20	10
Ref. 36	Vitamin B12	5 x 10^{14}	4	No	14	0.09
Ref. 30	Citrulline	~10^{15}	100	Yes (1)	16	65
Ref. 25	Theophyll-ine	~10^{14}	50–100[f]	Yes (1)	15	0.1
Ref. 31	Valinamide	3 x 10^{13}	<10	Yes (1)	25	3000
Ref. 29	Guanosine	5 x 10^{13}	<20[f]	Yes (1)	14	30
Ref. 29	Guanosine/ Arginine	5 x 10^{13}	30–50[f]	Yes (1)	14	200/4000
Ref. 37	FAD	~10^{15}	10^5	Yes (2)	11/13	1
Ref. 37	FMN	~10^{15}	200	Yes (1)	11	0.5
Ref. 38	NMN	5 x 10^{14}	2	No	10–20	2.5
Ref. 46	Biphenyl T.S. Analog	4 x 10^{14}	3	No	≤25	7
DNA						
Ref. 39	Organic					
Ref. 40	ATP	2 x 10^{14}	10^7–10^8	No	15	10

[a] Does not necessarily reflect primary sequence conservation.
[b] Only Cibacron Blue and Reactive Blue 4 aptamers are reported.
[c] These values are based on enrichment numbers. Sequencing suggested the complexity was actually lower
[d] Many of these bases were forced to be constant due to base pairing with the defined primer binding sites.
[e] Not including base paired regions.
[f] Estimate; enrichment data are reported.

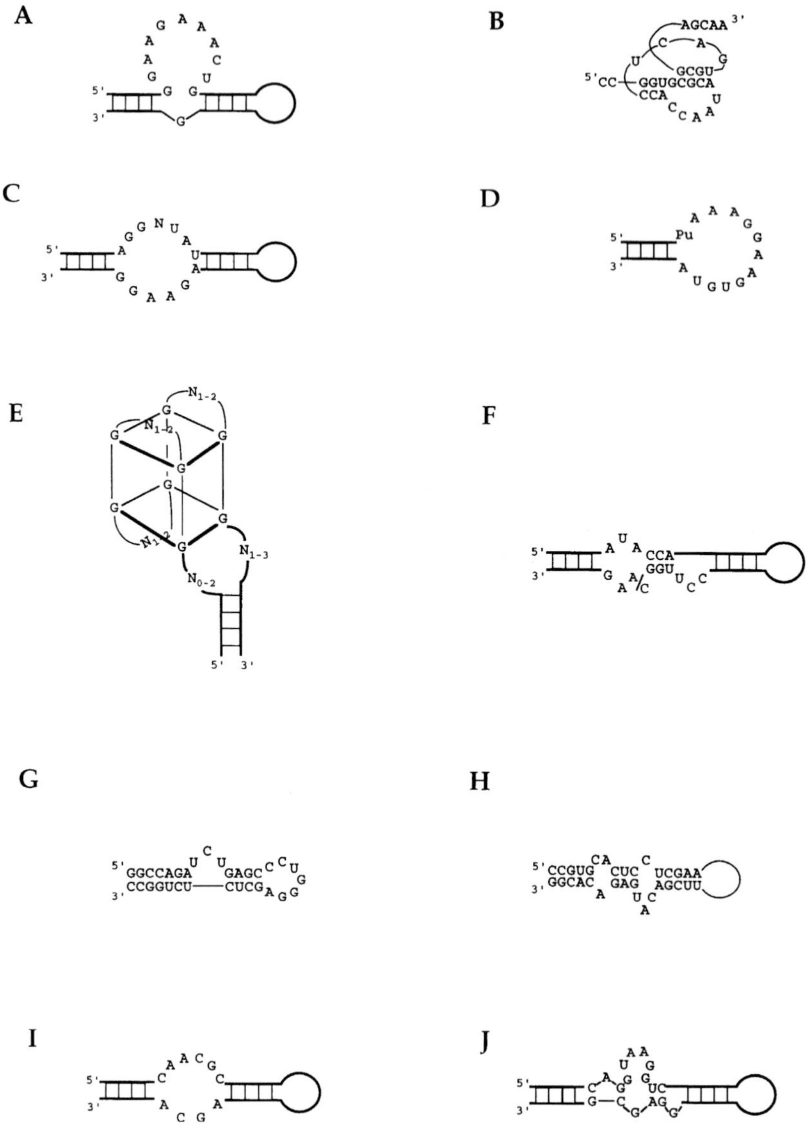

Figure 4.2: A rogues gallery of RNA aptamers. Proposed secondary structures for :

a) ATP aptamer; (Famulok and Szostak, 1992)
b) cyanocobalamin aptamer; (Wilson and Szostak, in press)
c) FMN aptamer; (Lorsch and Szostak, 1994)
d) FAD aptamer; (Lorsch and Szostak, 1994)
e) riboflavin aptamer; (Burgstaller and Famulok, 1994)
f) theophylline aptamer; (Ellington and Szostak, 1990)
g) HIV-1 Tat RNA, arginine binding motif; (Tan et al., 1993)
h) arginine aptamer; (Sassanfor and Szostak, 1993)
i) arginine-guanosine joint aptamer; (Connell et al., 1993)
j) arginine aptamer; (Conell et al., 1994)

diverse. A great deal should be learned about how RNA and DNA specifically recognize small molecules, and for that matter large molecules, as NMR and crystallographic data become available for *in vitro* selected receptors.

One feature that does seem to be common to most if not all aptamers is conformational change upon ligand binding. All RNA aptamers to small molecules for which structural information is available appear to undergo some conformational change upon ligand binding. The ATP aptamer, (Famulok and Szostak, 1992) the cyanocobalamin aptamer, (Wilson and Szostak, in press) and the guanosine/arginine joint aptamer (Connell et al., 1993) all have bases that become hypersensitive to chemical modification in the presence of ligand. These data strongly suggest that the RNAs undergo structural changes when they bind their ligands, and that these structural changes cause the bases in question to become hyperexposed. NMR spectra of the theophylline aptamer show major changes when theophylline is present relative to when it is absent, leading Jenison, et al. to conclude that this aptamer, as well, changes conformation when it binds its ligand (Ellington and Szostak, 1990). It is worthy of note that RNA binding proteins also induce large structural changes in their cognate RNAs upon binding (Davies et al., 1993). X-ray crystallography has shown that $tRNA^{asp}$ changes structure greatly when it is bound to aspartyl tRNA synthetases (Huizenga and Szostak, 1995; Rould et al., 1991). NMR analysis has shown that both the RRE and TAR RNAs from HIV change conformation when bound by the Rev and Tat proteins, respectively (Tan et al., 1993; Cavarelli et al., 1993; Battiste et al., 1994). It may be that these sorts of structural changes will be a common feature in RNA recognition.

4.9 Transition State Stabilization and Tight Binding

As soon as it was demostrated that RNA could specifically bind small molecules, it was proposed that in vitro selection could be used to generate novel RNA-based catalysts by isolating RNAs that bound to transition state analogues, and stabilized the transition state of a reaction relative to the ground states. (Tuerk and Gold, 1990) This technique had already been used to great effect by researchers working with monoclonal antibodies. (Petersen et al., 1994) Many groups have attempted to isolate such catalytic RNAs, but until recently they met with very limited success. RNAs were generated that bound transition state analogues with good affinity, but none of these aptamers proved to be catalytic. Presumably, these failures were due to the fact that the RNAs did not discriminate well enough between ground and transitions states. Because of these failures there was a good deal of pessimism as to whether RNA was a versitile enough receptor for catalysts to be generated in this way. Recently, however, Schultz and coworkers, using a prudent choice of transition state analogues, have shown that RNA aptamers can effect catalysis by specifically stabilizing a transition state. (Lerner et al., 1991) The reaction they chose to use is the isomerization of a sterically hindered bridged biphenyl. The transition state of the reaction will have the two phenyl rings in the same plane, and thus the transition state analogue used was a compound that had the two rings locked into planarity. Using *in vitro* selection they isolated three aptamers that specifically bind the transition state analogue. Of the three, one was shown to accelerate the isomerization reaction nearly 90-fold. The K_d of this aptamer for the transition state analogue is 7 μM, while the K_m for the substrate is 540 mM. The authors show that nearly all of the preferential binding energy for the transition state analogue relative to the substrate is realized in catalysis of the reaction. While the catalysis achieved by this RNA is modest, it is nearly certain that if tighter and more specific binders to the transition state analogue can be isolated, then better rate-enhancement will be achieved. It seems quite likely that the future will reveal many more catalytic aptamers for a wide variety of reactions.

As tight binding will be required for the generation of efficient catalytic aptamers, a brief discussion of possible methods of selecting for RNAs with low K_ds for their ligands is merited. The K_ds one can achieve with normal affinity column chromatography selections might be limited by the off rate of the RNA-ligand complex. If we assume a practical diffusion limit (Prudent et al., 1994) for k_{on} of 10^7 M^{-1} s^{-1} (although, given the fact that aptamers commonly undergo conformational changes upon ligand binding, this value may be too large), then the k_{off} for an aptamer with a K_d of 1 nM should be 0.01 s^{-1}. In other words, it will take nearly two minutes for RNAs to be eluted from the matrix. So, unless one soaks the columns in affinity eluent, 1 nM is near the limit of K_ds that can be achieved using this technique. In practice K_ds this low have not yet been attained for aptamers for small molecules. There are a number of possible reasons for this failure. First, as mentioned above k_{on} may be lower than the diffusion limit due to rate-limiting conformational changes or low accessibility of the binding site. There are several possible ways to get around this potential problem. One could do the selection under competitive conditions, that is, have the RNA in excess over the number of ligand sites. By allowing the system to come to equilibrium it should be possible to select for aptamers with low off rates (and presumably, since RNAs with large k_{ons} will compete most effectively for the limited number of ligands, this will translate into lower K_ds). The use of cleavable linkers (e.g. disulfides, cis-diols, etc.) to attach the ligand to the column matrix could also allow the isolation of aptamers with low k_{offs}. It is also quite likely that tight binders are more complex structurally than weaker binders, and thus much less abundant in the initial random RNA or DNA library. Such complex aptamers might be isolated by *in vitro* evolution, i.e. the introduction of new mutations by mutagenic amplification during the process of continued selection under gradually more stringent selection conditions.

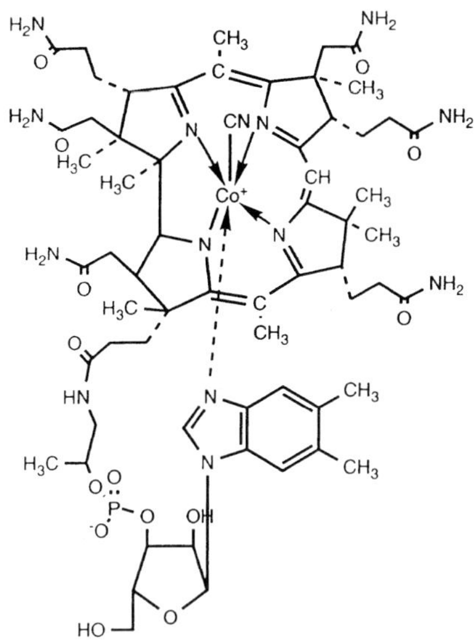

Cyanocobalamin

K_d = 90 nM

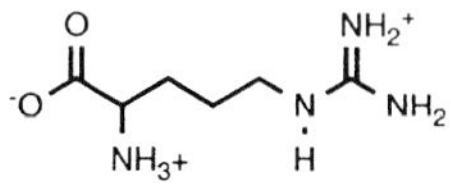

Arginine

K_d = 65-4 ,000 μM

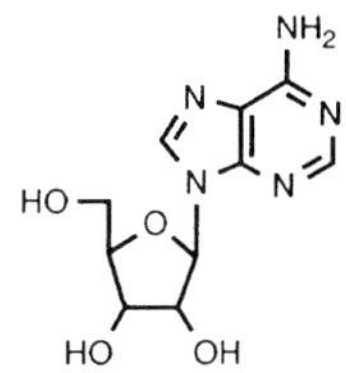

Adenosine

K_d = 0.7 μM

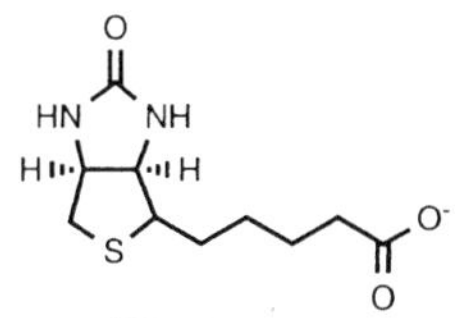

Biotin

K_d = 10 μM

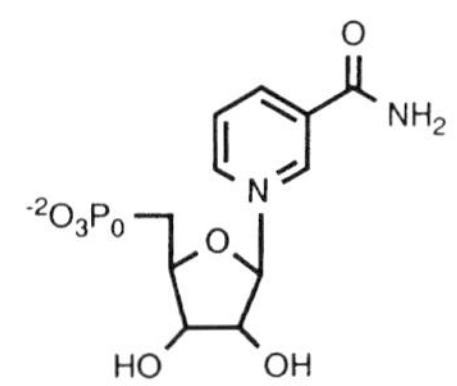

Nicotinamides

K_d = 2.5 μM

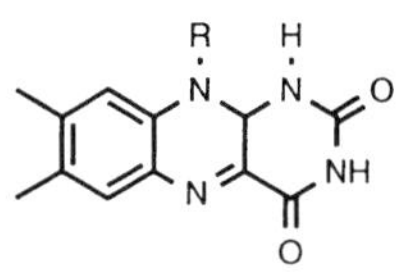

Flavins

K_d = 0.5 - 4 μM

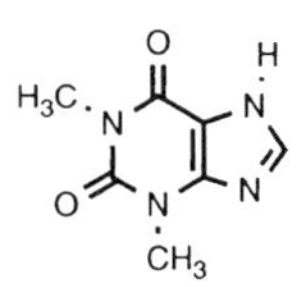

Theophylline

K_d = 100 nM

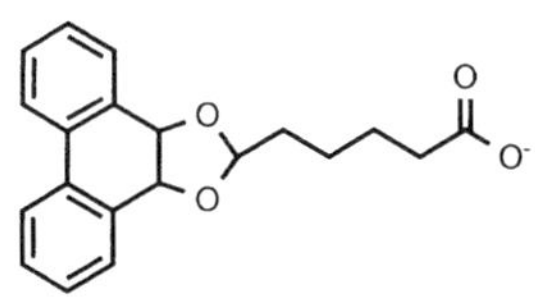

Bridged Biphenyl T.S. Analogue

K_d = 7 μM

Figure 4.3: Structures of some of the ligands for which aptamers have been isolated. The K_d(s) of the aptamers are shown below each molecule. (See text for references).

4.10 Conclusions

RNA and DNA are suprisingly robust receptors for small molecule ligands. *In vitro* selection has opened the way for the generation of useful and interesting aptamers, catalysts and other reagents. We expect that in the coming few years, *in vitro* selection will continue to be useful in unraveling the ways that nucleic acids interact with other molecules and effect catalysis. RNA aptamers have already proven to be useful starting points in the selection of novel ribozymes that utilize biologically relevent small-molecule substrates, (Benner et al., 1990; Lorsch and Szostak, 1994) and this is no doubt just a foreshadowing of things to come.

References

Ahsen, U.v., Davies, J., and Schroeder, R. (1991) Antibiotic Inhibition of Group I Ribozyme Function, Nature *353*, 368–370.

Ahsen, U.v. and Schroeder, R. (1991) Streptomycin Inhibits Splicing of Group I Introns by Competition with the Guanosine Substrate, Nuc. Acids Res. *19*, 2261–2265.

Bass, B.L. and Cech, T.R. (1984) Specific Interactions between the Self-splicing RNA of Tetrahymena and Its Guanosine Substrate: Implications for Biological Catalysis by RNA, Nature *308*, 820–826.

Bass, B.L. and Cech, T.R. (1986) Ribozyme Inhibitors: Deoxyguanosine and Dideoxyguanosine are Competitive Inhibitors of Self-splicing of the Tetrahymena Ribosomal Ribonucleic Acid Precursor, Biochemistry *25*, 4473–4477.

Battiste, J.L., Tan, R., Frankel, A.D., and Williamson, J.R. (1994) Binding of an HIV Rev Peptide to Rev Responsive Element RNA Induces Formation of Purine-Purine Base Pairs, Biochemistry *33*, 2742–2747.

Benner, S., Ellington, A.D., and Tauer, A. (1990) Modern Metabolism as a Palimpsest of the RNA World, Proc. Natn. Acad. Sci. USA *86*, 7054–7058.

Burgstaller, P. and Famulok, M. (1994) Isolation of RNA Aptamers for Biological Cofactors by In Vitro Selection, Angew. Chem. Int. Ed. Engl. *33*, 1084–1087.

Calnan, B.J., Biancalana, S., Hudson, D., and Frankel, A.D. (1991) Analysis of Arginine-rich Peptides from HIV Tat Protein Reveals Unusual Features of RNA-Protein Recognition, Genes Dev. *5*, 201–210.

Calnan, B.J., Tidor, B., Biancalana, S., Hudson, D., and Frankel, A.D. (1991) Arginine-Mediated RNA Recognition: The Arginine Fork, Science *252*, 1167–1171.

Cavarelli, J., Rees, B., Ruff, M., Thierry, J.-C., and Moras, D. (1993) Yeast $tRNA^{Asp}$ Recognition by Its Cognate Class II Aminoacyl-tRNA Synthetase, Nature *362*, 181–184.

Connell, G.J., Illangesekare, M., and Yarus, M. (1993) Three Small Ribooligonucelotides with Specific Arginine Sites Biochemistry *32*, 5497–5502.

Connell, G. and Yarus, M. (1994) RNAs with Dual Specificity and Dual RNAs with Similar Specificity, Science *264*, 1137–1141.

Davies, J., Ahsen, U.v., and Schroeder, R. (1993) Antibiotics and the RNA World: a Role for Low-molecular-weight Effectors in Biochemical Evolution, In: The RNA World, R.F. Gesteland and J.F. Atkins, Editor. (Cold Spring Harbor: Cold Spring Harbor Laboratory Press). p. 185–204.

Ellington, A.D. and Szostak, J.W. (1992) Selection In Vitro of Single-stranded DNA Molecules that Fold into Specific Ligand binding Structures, Nature *355*, 850–852.

Ellington, A.D. and Szostak, J.W. (1990) In Vitro Selection of RNA Molecules that Bind Specific Ligands, Nature *346*, 818–822.
Ellington, A.D., Personal Communication.
Famulok, M. and Szostak, J.W. (1992) Stereospecific Recognition of Tryptophan Agarose by In Vitro Selected RNA, J. Am. Chem. Soc. *114*, 3990–3991.
Famulok, M. (1994) Molecular Recognition of Amino Acids by RNA Aptamers: an L-Citrulline Binding RNA Motif and Its Evolution into an L-Arginine Binder, J. Am . Chem Soc. *116*, 1698–1706.
Famulok, M., Personal Communication.
Fersht, A. (1985) Enzyme Structure and Mechanism. 2nd ed. (New York, W.H. Freeman and Co.).
Herschlag, D. and Cech, T.R. (1990) Catalysis of RNA Cleavage by the Tetrahymena thermophila Ribozyme. 1. Kinetic Description of the Reaction of an RNA Substrate Complementary to the Active Site, Biochemistry *29*, 10159–10171.
Herschlag, D. and Cech, T.R. (1990) Catalysis of RNA Cleavage by the Tatrahymena thermophila Ribozyme. 2. Kinetic Description of the Reaction of an RNA Substrate That Forms a Mismatch at the Active Site, Biochemsitry *29*, 10172–10180.
Huizenga, D. and Szostak, J.W. A DNA Aptamer that Binds Adenosine and ATP, (1995). Biochemistry *34*, 656–665.
Jenison, R.D., Gill, S.C., Pardi, A., and Polisky, B. (1994) High-resolution Molecular Discrimination by RNA, Science *263*, 1425–1429.
Lauhon, C.T. and Szostak, J.W. RNA Aptamers that Bind Flavin and Nicotinamide Cofactors, (1995). Am. Chem. Soc. *117*, 1246–1257.
Lerner, R.A., Benkovic, S.J., and Schultz, P.G. (1991) At the Crossroads of Chemistry and Immunology: Catalytic Antibodies, Science *252*, 659–667.
Lorsch, J.R. and Szostak, J.W. (1994) In Vitro Evolution of New Ribozymes with Polynucleotide Kinase Activity, Nature *371*, 31–36.
Lorsch, J.R. and Szostak, J.W. (1994) In Vitro Selection of RNA Aptamers Specific for Cyanocobalamin, Biochemistry *33*, 973–982.
Majerfeld, I. and Yarus, M. (1994) An RNA Pocket for an Aliphatic Hydrophobe, Nat. Struct. Biol. *1*, 287–292.
Mattaj, I.W. (1993) RNA Recognition: A Family Matter, Cell *73*, 837–840.
Michel, F., Hanna, M., Green, R., Bartel, D.P., and Szostak, J.W. (1989) The Guanosine Binding Site of the Tetrahymena Ribozyme, Nature *342*, 391–395.
Moazed, D. and Noller, H.F. (1987) Interaction of Antibiotics with Functional Sites in 16S Ribosomal RNA, Nature *327*, (389–394).
Peterson, R.D., Bartel, D.P., Szostak, J.W., Horvath, S.J., and Feigon, J. (1994) 1H NMR Studies of the High-Affinity Rev Binding Site of the Rev Responsive Element of HIV-1 mRNA: Base Pairing in the Core Binding Element, Biochemistry *33*, 5357–5366.
Prudent, J.R., Uno, T., and Schultz, P.G. (1994) Expanding the Scope of RNA Catalysis, Science *264*, 1924–1927.
Puglisi, J.D., Tan, R., Calnan, B.J., Frankel, A.D., and Williamson, J.R. (1992) Conformation of the TAR RNA-Arginine Complex by NMR Spectroscopy, Science *257*, 76–80.
Rould, M.A., Perona, J.J., and Steitz, T.A. (1991) Structural Basis of Anticodon Loop Recognition by Glutaminyl-tRNA Synthetase, Nature *352*, 213–218.
Sassanfar, M. and Szostak, J.W. (1993) An RNA Motif that Binds ATP, Nature *364*, 550–553.
Tan, R., Chen, L., Buettner, J.A., Hudson, D., and Frankel, A.D. (1993) RNA Recognition by an Isolated Helix, Cell *73*, 1031–1040.
Tao, J. and Frankel, A.D. (1992) Specific Binding of Arginine to TAR RNA, Proc. Natl. Acad. Sci. USA *89*, 2723–2726.
Tuerk, C. and Gold, L. (1990) Systematic Evolution of Ligands by Exponential Enrichment: RNA Ligands to Bacteriophage T4 DNA Polymerase, Science *249*, 505–510.
Weeks, K.M., Ampe, C., Schultz, S.C., Steitz, T.A., and Crothers, D.M. (1990) Fragments of the HIV Tat Protein Specifically Bind TAR RNA, Science *249*, 1281–1285.

Wilson, C. and Szostak, J.W. In Vitro Evolution of a Self alkylating Ribozyme, Nature, In Press.
Yarus, M. (1988) A Specific Amino Acid Binding Site Composed of RNA, Science *240*, 1751–1758.
Yarus, M. (1989) Specificity of Arginine Binding by the Tetrahymena Intron, Biochemistry *28*, 980–988.
Yarus, M. (1993) An RNA-Amino Acid Affinity, In: The RNA World, R.F. Gesteland and J.F. Atkins, Editor. (Cold Spring Harbor: Cold Spring Harbor Laboratory Press). p. 205–217.
Yarus, M. and Majerfeld, I. (1992) Co-optimization of Ribozyme Substrate Stacking and L-arginine Binding, J. Mol. Biol. *225*, 945–949.
Zapp, M.L., Stern, S., and Green, M.R. (1993) Small Molecules that Selectively Block RNA Binding of HIV-1 Rev Protein Inhibit Rev Function and Viral Production, Cell *74*, 969–978.

5 Discovery and Characterization of a Thrombin Aptamer Selected from a Combinatorial ssDNA Library

Linda C. Griffin, Lawrence L.K. Leung

5.1 Introduction

Aptamer technology combines the capacity for enormous structural diversity in random pools of oligonucleotides with the power of the polymerase chain reaction (PCR). In general, this technology involves the screening of large, random-sequence pools of RNA and is based on the premise that the random oligonucleotides assume a large number of tertiary structures, some of which may possess desirable binding or catalytic activity against target molecules (Ellington and Szostak, 1990). The screening involves the use of chromatography techniques that allow separation of the oligonucleotides with the highest affinity for the target from the rest of the pool. The selected oligonucleotides can then be amplified by PCR to give a new pool enriched in sequences with affinity for the target. Additional rounds of the selection and amplification cycle allow the isolation of a product pool containing only those oligonucleotides, termed "aptamer" (from the Latin "aptus" meaning to fit) (Ellington and Szostak, 1990) having the greatest affinity. Individual aptamer sequences from this pool can then be identified through cloning. The striking advantage that this technique provides over classic large-scale natural product screening is that a far larger population of molecules ($>10^{13}$) can be rapidly screened.

Because our primary interest is the development of nucleotide-based therapeutics (Bischofberger and Shea, 1992), we were interested in whether related selection techniques could be successfully applied to important therapeutic targets, including those not known to interact physiologically with nucleic acid, such as thrombin. Thrombin is a serine protease responsible for platelet activation, the conversion of fibrinogen to fibrin and the activation of coagulation factors V, VIII, XI and XIII (Fenton, 1981; Shuman, 1986; Furie and Furie, 1988; Mann et al., 1990; Gailani and Broze, 1991). Because of its pivotal role in both hemostasis and thrombosis, thrombin is a major target for the development of antithrombotic therapeutics.

RNA and DNA are chemically similar, however, ssDNA is not characterized as possessing the structural diversity of RNA inasmuch as DNA sequences that are structurally and functionally analogous to ribosomal RNAs, transfer RNAs, snRNAs or ribozymes have not been observed in nature. Nevertheless, because of the desirable property of being more stable *in vivo* than RNA, we were also interested in determining whether ssDNA could be used to obtain aptamers. The result of applying ssDNA aptamer technology to a target protein not known to interact physiologically with nucleic acid was the identification of a ssDNA 15-mer, with a unique tertiary structure, capable of nanomolar inhibition of thrombin activity *in vitro* and potent anticoagulant activity *in vivo*.

5.2 Discovery and Initial Characterization

5.2.1 Selection Method

First, a pool of greater than 10^{13} unique ssDNA 96-mers, containing a 60-nucleotide random region flanked by 18-nucleotide PCR primer binding sites, was chemically synthesized. This pool was amplified by PCR in the presence of radiolabeled nucleotide triphosphates and a 5′-biotinylated primer for one strand. The desired ssDNA was isolated from its biotinylated complementary strand after application to an avidin-agarose column and base denaturation (Hultman et al.,

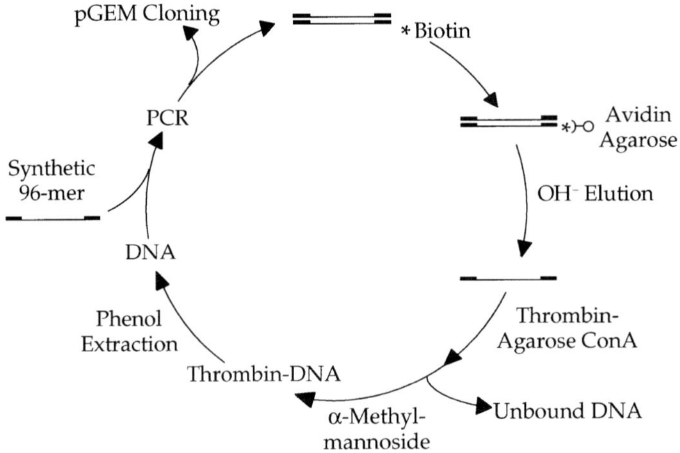

Figure 5.1. Schematic diagram for the selection of DNA aptamers to human thrombin.

1988). To remove DNA with affinity for the chromatography matrix, the ssDNA was first applied to a concanavalin A (ConA) -agarose column and then the eluent was applied to human thrombin immobilized on ConA. The unbound DNA was washed from the column and thrombin-DNA complexes were eluted with a ConA ligand, α-methylmannoside. The DNA from those fractions containing thrombin was isolated, quantitated and amplified by PCR. Additional rounds of the selection and amplification cycle were performed, each time enriching the DNA pool for thrombin binders, until no further increase in the percent of DNA that eluted with thrombin was observed (Figure 5.1).

In the first round of selection only 0.01% of the input DNA eluted with thrombin; this percentage increased to ~40% by the fifth round. DNA from the fifth round was assayed for thrombin binding specificity on nitrocellulose filters. The aptamer DNA bound thrombin and to a lesser degree prothrombin, the inactive precursor to thrombin. The aptamer DNA did not bind serine proteases trypsin, chymotrypsin or tissue plasminogen activator, or the control proteins, albumin and kallikrein. Filter binding studies indicated that the stoichiometry of aptamer binding to thrombin was 1:1. The pool of DNA used in the first round of selection showed very little affinity for thrombin or any of the other proteins. Thus, from a large random pool of ssDNA we isolated a subset of ssDNA molecules with increased affinity and specificity for thrombin (Bock et al., 1992).

Thrombin is a glycoprotein and we were therefore able to use lectin-agarose for thrombin immobilization and α-methylmannoside to elute the thrombin-DNA complexes. Once separated from the matrix, DNA-thrombin complexes were denatured and the thrombin binders isolated. Initially, we applied ssDNA to thrombin that was covalently linked to agarose. The isolation of thrombin binders required the denaturation of the thrombin-aptamer complexes while still bound to the matrix. In addition to thrombin aptamers, denaturing elution with EDTA gave ssDNA with affinity for the matrix and hence only a modest enrichment of thrombin aptamers. These results suggested that the conditions under which aptamers are eluted from a covalently bound target are critical to the successful isolation of high affinity aptamers. One way to circumvent such problems is to use a reversible linker that permits mild elution of aptamer-target complexes from the matrix. After separation from matrix associated oligonucleotides, the complexes can be fully denatured and the aptamers with the highest affinity recovered. Our experience with thrombin selections suggests that lectin immobilization may provide an aptamer selection technique applicable to a large variety of glycoproteins. Other methods for non glycoprotein targets that allow for the specific elution of aptamer-target complexes from a matrix should also be advantageous.

5.2.2 Sequence Analysis

Having isolated a subset of ssDNA molecules with increased affinity and specificity for thrombin the next step was to identify the sequence(s) responsible for conferring thrombin affinity on the 96-mer ssDNA aptamers. We determined the nucleotide sequence of the 60-nucleotide randomly generated region from 32 clones. Sequence analysis showed each of the 32 clones to be different, however, a striking sequence conservation was evident in every clone. The hexamer 5′-GGTTGG appeared at a variable location within the 60-nucleotide randomized region in 31 out of 32 clones. Moreover, in 28 of the 32 clones, a second hexamer 5′-GGNTGG was located 2–5-nucleotides 5′ or 3′ from the hexamer 5′-GGTTGG. Except for four clones that presented a close variation, the sequence 5′-GGNTGGN$_{(2-5)}$GGNTGG was conserved (Figure 5.2). DNA sequencing of clones from the unselected DNA pool and from a pool of aptamers selected for binding

Clone																				
1		g	G	G	t	T	G	G	-	-	g	t	c	G	G	t	T	G	G	t
2		g	G	G	a	T	G	G	-	-	t	t	t	G	G	t	T	G	G	g
3		a	G	G	t	T	G	G	-	-	-	g	a	G	G	g	T	G	G	g
4		t	G	G	t	T	G	G	-	-	c	g	a	G	G	a	T	G	G	a
5		a	G	G	t	T	G	G	-	g	t	a	g	t	G	t	T	G	G	t
6		a	G	G	t	T	G	G	-	-	g	c	t	G	G	t	T	G	G	g
7		g	G	G	t	T	G	G	-	-	-	g	a	G	G	t	T	G	G	a
8		t	G	G	t	T	G	G	-	-	g	t	c	G	G	t	T	G	G	g
9		g	G	G	a	T	G	G	-	-	t	g	t	G	G	t	T	G	G	c
10		t	G	G	t	T	G	G	-	c	g	a	g	G	G	a	T	G	G	g
11		t	G	G	a	T	G	G	-	-	t	g	a	G	G	t	T	G	G	a
12		g	G	G	g	T	G	G	-	-	t	t	a	G	G	t	T	G	G	t
13		a	G	G	g	T	G	G	-	-	t	t	a	G	G	t	T	G	G	t
14		c	G	G	t	T	G	G	-	g	t	t	g	G	G	a	T	G	G	a
15		c	G	G	t	T	G	G	-	-	t	g	t	G	G	t	T	G	G	t
16		a	G	G	t	T	G	G	-	-	t	g	t	G	G	g	T	G	G	g
17		c	G	G	g	T	G	G	-	-	a	t	a	G	G	t	T	G	G	a
18		g	G	G	g	T	G	G	-	-	t	g	t	G	G	9	T	G	G	g
19		t	G	G	t	T	G	G	t	t	a	c	t	G	G	t	T	G	G	g
20		g	G	G	t	T	G	G	-	-	t	c	t	G	G	g	T	G	G	a
21		t	G	G	t	T	G	G	-	-	g	t	t	G	G	g	T	G	G	a
22		t	G	G	t	T	G	G	-	-	c	c	a	G	G	t	T	G	G	a
23		t	G	G	t	T	G	G	g	t	6	c	a	G	G	t	T	G	G	g
24		t	G	G	g	T	G	G	-	-	g	g	a	G	G	t	T	G	G	t
25		a	G	G	t	T	G	G	-	-	t	t	t	G	G	g	T	G	G	t
26		a	G	G	t	T	G	G	-	t	t	a	g	G	G	t	T	G	G	t
27		g	G	G	a	T	G	C	-	-	g	g	t	G	G	t	T	G	G	g
28		t	G	G	t	T	G	G	-	t	t	a	t	G	G	t	T	G	G	t
29		a	G	G	t	T	G	G	-	-	t	g	t	G	G	t	T	G	G	c
30		a	G	G	t	T	G	G	-	-	t	g	t	G	G	g	T	G	G	g
31		t	G	G	t	T	G	G	-	-	-	g	a	G	G	t	T	G	G	t
32		g	G	G	t	T	G	G	t	g	g	g	t	G	G	A	T	G	G	t
	G		32	32	5	0	32	31						31	32	7	0	32	32	
	A		0	0	4	0	0	0						0	0	4	0	0	0	
	T		0	0	23	32	0	0						1	0	21	32	0	0	
	C		0	0	0	0	0	1						0	0	0	0	0	0	
			G	G	t	T	G	G			(N)3			G	G	t	T	G	G	

Figure 5.2: Thrombin aptamer sequence homology. The region of homology identified is shown in bold upper case. Below is a tabulation of the data and the deduced consensus sequence.

to a different target revealed no homology to the thrombin-selected aptamers. The strong sequence conservation in the thrombin aptamers suggested that this "consensus sequence" was responsible, entirely or in part, for conferring thrombin affinity on the 96-mer oligonucleotides (Bock et al., 1992).

5.2.3 Inhibition of Clotting

The thrombin aptamer selection was based solely on the ability to bind thrombin. To characterize the aptamers functional properties we analyzed whether the aptamer DNA could inhibit the thrombin-catalyzed conversion of fibrinogen to fibrin as measured by a prolongation of thrombin clotting times (Table 5.1).

Tabelle 5.1. Sequence specificity of thrombin inhibition.

Sample	Sequence	Conc.	Clotting times (sec)[1] Purified fibrinogen	Human plasma
No DNA	---	---	25 ± 1	25 ± 1
Clone 29	96-mer	100 nM	76 ± 3	50 ± 2
Unselected DNA	96-mer	100 nM	26 ± 1	26 ± 1
Consensus 15-mer	GGTTGGTGTGGTTGG	100 nM	169 ± 8	43 ± 2
Scrambled 15-mer	GGTGGTGGTTGTGGT	100 nM	26 ± 1	26 ± 1
Consensus 6mer	GGTTGG	20 mM	42 ± 2	40 ± 2
Scambled 6-mer	TGGGGT	20 mM	26 ± 1	26 ± 1

[1]Thrombin (13 nM final conc.) in selection buffer (20 mM Tris-acetate, pH 7.4, 140 mM NaCl, 5 mM KCl, 1 mM $MgCl_2$) was pre-equilibrated for 1 min at 37 °C. DNA samples were incubated for 1 min at 37 °C in either selection buffer containing human fibrinogen (2 mg/mL) or fresh human plasma and added to the thrombin to give the indicated final concentrations of DNA. Clotting times were measured using an aut9mated fibrometer.

Using a 96-mer oligonucleotide which contained the sequence, 5′-GGTT GGTGTGGTTGG, and purified fibrinogen, fibrin formation by thrombin was significantly delayed compared to the unselected DNA control. More impressively, with respect to oligonucleotide size and potency, a synthetic 15-mer consisting of the consensus sequence 5′-GGTTGGTGTGGTTGG was even more potent. The 6-mer 5′-GGTTGG also inhibited thrombin, although a higher concentration was required to detect inhibition. Subsequent studies examining the dependence of inhibition on aptamer concentration indicated that the 15-mer inhibited thrombin as a monomer while the 6-mer inhibited thrombin as a dimer. No inhibition was detected with 15-mer or 6-mer scrambled sequence controls. The aptamers lost some potency in human plasma, which may be due to binding of prothrombin or other plasma proteins present in much higher concentrations than thrombin (Shaw et al., 1995). These results demonstrated that not only were the thrombin aptamers functionally active but also that this activity was sequence dependent.

Moreover, this activity was retained by a short consensus sequence derived from the selected 96-mers (Bock et al., 1992).

Of note, the aptamers did not inhibit the amidolytic activity of thrombin towards a small chromogenic peptide substrate S-2238. Similar values for k_{cat} and K_m were obtained when assays were performed with or without oligonucleotides, suggesting that the structure of the active site was not significantly altered (Wu et al., 1992) and that the thrombin aptamer inhibited thrombin activity by interfering with substrate binding outside the active site.

5.3 Structure

5.3.1 Nuclear Magnetic Resonance Spectroscopy (NMR) studies

The potent inhibitory activity of the thrombin aptamer, 5′-GGTTGGTGTGG TTGG, in the thrombin clotting assay, indicated high thrombin affinity and thermal stability. This suggested that the 15-mer was likely to be an ordered structure conducive to NMR studies. The composition of the aptamer being only G and T residues rules out any Watson-Crick base pairing and, since G-T wobble base pairs do not offer high thermal stability, their presence was considered unlikely. It was found by NMR that the aptamer adopts a tertiary structure which is unlike any structure previously documented for DNA (Wang et al., 1993a; Wang et al., 1993b; Macaya et al., 1993).

The structural characterization of the aptamer began by obtaining the imino proton spectrum to determine the extent of base pairing. The preliminary characterization of the aptamer by NMR gave variable results; however, reproducible results were obtained with the use of buffer which contained 5 mM KCl. The imino proton spectrum showed the presence of 8 imino protons which were slowly (hours) exchanging with water and examination of the imino proton region as a function of temperature showed that the aptamer melted at 50 °C. This kinetic and thermodynamic stability of the aptamer suggested that there were no large structural changes upon binding to thrombin. The one-dimensional proton NMR spectrum also showed that only one form of the aptamer was present in solution and the observed chemical shifts were suggestive of an unusual structure (Guschlbauer et al., 1990).

Two-dimensional NMR techniques that indicate through space and through bond connectivities between the protons was used to assign the proton spectrum. The imino spectrum and the proton connectivities of the imino protons suggested the presence of G-quartets. Connectivities involving H^8 of G suggested also that four of the G residues were in syn-conformation. In addition, several proton connectivities between non-adjacent residues were observed. Only one folding

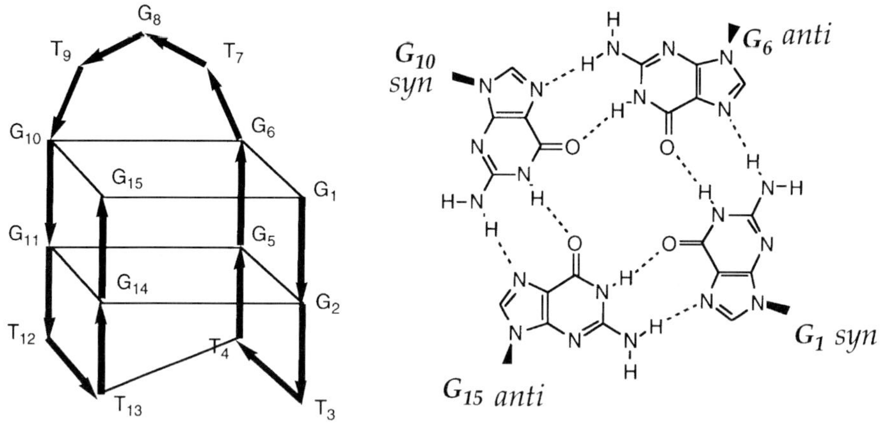

Figure 5.3 A: A depiction of the tertiary structure of the thrombin aptamer. The residues G_1, G_5, G_{10}, and G_{14} are in a syn conformation with respect to the glycosidic bond. The thin lines indicate the location of the G-quartets (G_1-G_6-G_{10}-G_{15} and G_2-G_5-G_{11}-G_{14}) and the direction of the backbone is indicated by the arrows.
B: Syn-anti-syn-anti G-quartet.

pattern (Figure 5.3 A, B) was found to be consistent with all of the nuclear Overhauser effect connectivities.

The thrombin aptamer structure is highly compact, with two G-quartets which are in syn-anti-syn-anti conformation with respect to the glycosidic bonds, a loop which is partially folded over one quartet and two T-T loops folded under the other quartet. Structures (Figure 5.4 A and 4 B) were obtained from the combined use of NMR NOE, scalar coupling data and molecular modeling. This model is consistent with all the NMR results, as well as the consensus sequence data.

Some of the same structural features observed for the intramolecular quartets of the aptamer in solution have previously been reported in telomeres where G-quartets form between separate DNA strands (Hardin et al., 1991; Zahler et al., 1991). The stabilization that K^+ provides the aptamer structure is consistent with data showing K^+ to be important in the stabilization of some tetraplex structures (Guschlbauer et al., 1990). It has been proposed that the presence of tetraplex structures in telomeres may protect the ends of chromosomes (Henderson et al., 1987) as well as playing a role in meiosis (Liu and Gilbert, 1994). It was unexpected that telomere DNA and a ssDNA thrombin aptamer would share similar structural features. This finding suggests that the G-quartet may be a basic structural element of ssDNA and found in aptamers to targets other than thrombin, and perhaps, in naturally occurring systems other than telomeres.

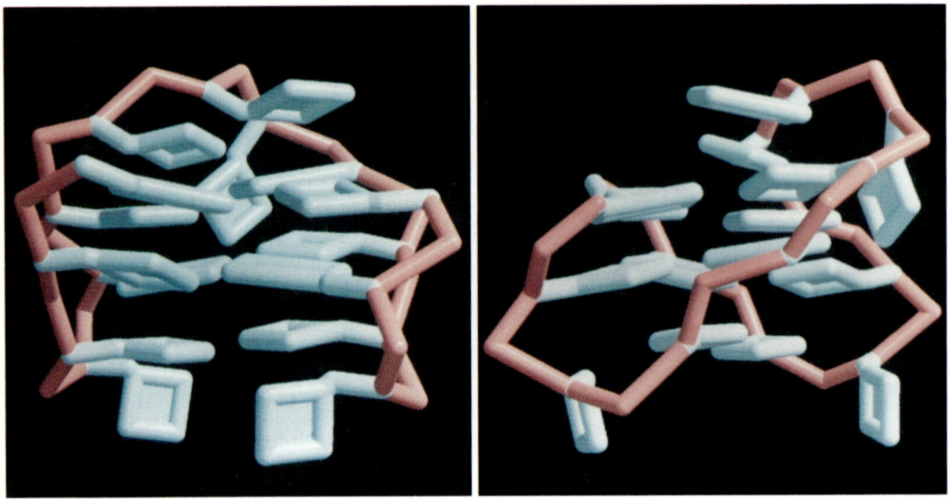

Figure 5.4: The structure of the aptamer is shown with the bases in blue and the backbone in pink. In panel A the "front" view of the aptamer is shown with residue 1 in the upper right and residue 15 in the upper left and the TGT loop on top. This view illustrates the compact nature of the aptamer structure and the orientation of the T_7-G_8-T_9 loop relative to the G_1-G_6-G_{10}-G_{15} quartet. In panel B a "side" view of the aptamer is shown with residue 1 in front of residue 15. This view illustrates the helicity of the aptamer backbone and that the backbone contains essentially four straight segments connected by three loops.

5.3.2 Structure-Activity Relationship Studies

The thrombin aptamer tertiary structure, unlike the duplex structure of B-DNA, is highly sequence specific. If the aptamer tertiary structure is important for thrombin binding and inhibition, activity should correlate with having a sequence compatible with the tertiary structure of the aptamer. To further study the structure-activity relationship, a series of singly substituted modified aptamers were synthesized and the concentrations of these compounds which gave rise to a doubling of the thrombin clotting time were compared to the unmodified aptamer (Wang et al., 1993; Krawczyk et al., 1995).

Before examining the activity of modified aptamers we wished to determine whether T_7-G_8-T_9, which was chosen on limited sequencing data (see Figure 5.2), was the best choice for the trinucleotide loop. To avoid having to synthesize and test all 64 possible trinucleotide combinations, we prepared four mixed pools where one position was fixed and the other two were random (GGTTGG NNXGGTTGG, N = G+A+T+C, X = G, A, T or C). The activity of the pool GGTTGGNNTGGTTGG was determined to be superior to the other pools. Thus the composition of the next four pools was GGTTGGXNTGGTTGG and the

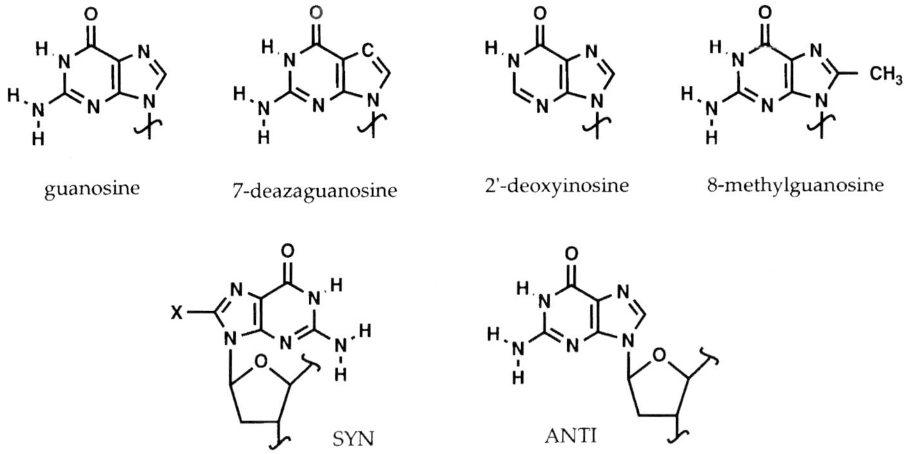

Figure 5.5: Single base substitutions. Syn and anti glycosidic bond orientations showing the favored syn orientation for 8-substituted guanosine and anti orientation for unsubstituted guanosine.

activity of the GGTTGGTNTGGTTGG pool was found to be greatest. Finally, we prepared the sequences GGTTGGTXTGGTTGG and determined that the sequence GGTTGGTGTGGTTGG had the highest activity and that we had initially chosen the best trinucleotide loop. The difference between the best trinucleotide combination and the worst was a factor of ~10 in potency. This difference indicated that the trinucleotide loop between the GGTTGG hexamers was more than just a spacing element in the structure.

Having determined that GGTTGGTGTGGTTGG was likely the most active 15-mer thrombin aptamer, we proceeded to make deletions from the 5′ and 3′ ends. Single deletions gave the striking result of reducing the activity 100-fold to that of the 6-mer, GGTTGG. The loss of one G on either end of the 15-mer disrupts the ability of the molecule to form a required G-quartet.

Data from single abasic sugar (Figure 5.5) substitutions at each position in the 15-mer was consistent with the consensus sequence (Table 5.2). Substitution at each conserved position (uppercase), GGtTGGtgtGGtTGG, resulted in a 100-fold loss of activity. The conserved Gs are each required for G-quartet formation and the NMR structure shows the two conserved T bases (T_4 and T_{13}) to be involved in hydrogen bonding interactions with each other and base stacking interactions with the G-quartets. Single and multiple abasic substitutions at the five non conserved positions were generally tolerated; substitution at T_7 was actually slightly better than the natural 15-mer. In the model of the 15-mer, T_7 projects away from the compact G-quartet structure while the other two bases in the loop stack on top of each other over one G-quartet. The data support that the T_7 base is not required for folding of the structure or interaction with thrombin. The abasic substitutions were completely consistent with the structure proposed for the aptamer since substitution for any of the residues involved in either G-quartet or stacking on a G-quartet is expected to disrupt the structure and hence binding to thrombin.

Table 5.2: Effect of single modified base substitutions on the thrombin aptamer concentration which doubles the thrombin clotting time.

	abasic	7-deazaG	inosine	8-methylG
G1	> 2 µM	> 2 µM	> 2 µM	~ 20 nM
G2	> 2 µM	> 2 µM	> 2 µM	~ 160 nM
T3	~75 nM			
T4	> 2 µM			
G5	> 2 µM	> 2 µM	> 2 µM	~ 20 nM
G6	> 2 µM	> 2 µM	> 2 µM	~ 80 nM
T7	~ 15 nM			
G8	~ 120 nM	~ 50 nM	~ 40 nM	~ 25 nM
T9	> 150 nM			
G10	> 2 µM	> 2 µM	> 2 µM	~ 20 nM
G11	> 2 µM	> 2 µM	> 2 µM	~ 80 nM
T12	> 75 nM			
T13	> 2 µM			
G14	> 2 µM	> 2 µM	> 2 µM	~ 20 nM
G15	> 2 µM	> 2 µM	> 2 µM	~ 65 nM

All of the G residue positions were individually substituted with 7-deazaguanosine and with inosine (Figure 5). 7-Deazaguanosine lacks the N7 of guanosine, which is the hydrogen bond acceptor of an amino group and inosine lacks the amino group which is involved as a hydrogen bond donor to an N7 in the G- quartets. Substitution of either 7-deazaguanosine or inosine for any of the G residues involved in the G-quartets significantly reduced activity (Table 5.2). Substitution of G_8, which is the only G not involved in a G quartet, had a minor effect as evidenced by the approximately 2-fold increase in concentration required to double the thrombin clotting time. The structure's G residues 1, 5, 10, and 14 have syn orientations about the glycosidic bond while all other G residues have anti orientations. Due to steric reasons, 8-methylguanosine has a preference for the syn orientation, and substitution of 8-methylguanosine for any of the syn G residues had little or no impact on thrombin inhibition. As expected on the basis of the structure, it was found that substitution of 8-methylguanosine for any of the anti G residues, which were involved in a G-quartet, led to a decrease in inhibitory activity. Substitution for G_8 had little or no effect on the thrombin inhibitory activity.

Thus, these structural modification studies corroborated and confirmed the NMR structure and indicate that this structure must be preserved for efficient thrombin inhibition. Additional modifications that introduce hydrophobic groups and replace phosphates in the backbone without disrupting G-quartet formation have been found to increase the thrombin inhibitory potency *in vitro* and *in vivo*.

5.4 Binding Site on Thrombin

5.4.1 Competition Studies

Unlike active site thrombin inhibitors, the thrombin aptamer does not inhibit the cleavage of small chromogenic amide substrates (Wu et al., 1992; Bock, personal communication) and therefore does not inhibit by binding directly to the active site of thrombin. However, distinct from the active site of thombin are two highly basic regions that form secondary binding sites on the surface of the molecule. These sites are important for thrombin's specificity in interactions with several physiologic macromolecular substrates and receptors. One site, the fibrinogen-binding exosite, contributes to the formation of a tight specific complex with fibrinogen (Fenton et al., 1988; Church et al., 1989), thrombomodulin (Wu et al., 1991), the platelet thrombin receptor (Vu et al., 1991), heparin cofactor II (Sheehan et al., 1991; Rogers et al., 1992) and the leech thrombin inhibitor hirudin (Chang, 1989; Grutter et al., 1990). The second site, the heparin recognition site, contributes to the significant increase of thrombin inactivation by antithrombin III in the presence of heparin (Chang et al., 1979; Messmore et al., 1979). Because of the polyanionic nature of the nucleic acid phosphodiester backbone, these basic sites on thrombin, shown to interact with acidic regions of other thrombin-binding molecules, are likely targets for thrombin aptamer binding.

To characterize the thrombin-aptamer interaction, a solid-phase binding assay was developed in which binding of soluble biotinylated thrombin aptamer to immobilized thrombin was quantified (Paborsky et al., 1993). Gamma-thrombin, an autoproteolyzed derivative of thrombin which contains a cleavage in the fibrinogen-binding exosite and two other sites, was assayed for its ability to compete with immobilized α-thrombin for aptamer binding. Soluble α-thrombin effectively competed with the immobilized α-thrombin while γ-thrombin, even at a 2000-fold excess concentration, did not compete with immobilized α-thrombin for aptamer binding. This result indicated that the regions that differ between α-thrombin and γ–thrombin may participate in thrombin aptamer binding. Furthermore, hirudin, a potent and specific thrombin fibrinogen binding exosite inhibitor, effectively competed with the aptamer for thrombin binding. In addition, it was shown in competition equilibrium binding experiments that binding of labeled thrombin to thrombomodulin was inhibited by the thrombin aptamer (Wu et al., 1992). Taken together, these data suggested that the fibrinogen-binding exosite is important for aptamer-thrombin interactions.

5.4.2 Lysine Protection Assay

To identify the essential lysine residues in human α-thrombin that participate in the aptamer-thrombin interaction, both the thrombin-aptamer complex and thrombin alone were derivatized with fluorescein 5′ isothiocyanate (FITC), a reagent for ε-amino group modification (Paborsky et al., 1993). The basic premise of these experiments is that in the aptamer-thrombin complex, specific lysine residues in the thrombin binding site are shielded and become relatively inaccessible to FITC modification. Following the modification by FITC, the thrombin-aptamer complex and thrombin alone were reduced, carboxymethylated and digested with endoproteinase Arg-C. The digestion products were analyzed by reversed-phase HPLC. In the absence of the aptamer, approximately twenty FITC-labeled peaks were generated. As a result of aptamer binding, the signal from 4 peaks was consistently reduced. The peptides from these peaks, P1–4, were collected and subjected to N-terminal sequence analysis. Peptides P1, P2 were of identical sequence and demonstrated that the lysine at position 65 was protected from derivatization by FITC. Sequencing of P3 and P4 indicated that the aptamer also protects the lysine at position 21 from FITC modification. Peptides P1 and P2 have identical amino acid composition but different retention times as did peptides P3 and P4. This was probably due to a rearrangement of a small portion of the fluorescein group which alters the hydrophobic properties of the peptides and hence their retention times in the C18 column (Mitchinson et al., 1982; Farley et al., 1984; Phillips, 1988).

These results indicated that both Lys21 and Lys65, which are in the fibrinogen-binding exosite of a-thrombin, are located within or in close proximity to the thrombin aptamer binding site. Lys21 and Lys65 also participate in fibrinogen-thrombin and hirudin-thrombin interactions (Church et al., 1989; Chang, 1989). The possibility exists that the binding of ssDNA induces conformational changes that interfere with the chemical modification of lysine residues that are not directly located in the combining site. However, the lack of competition by γ-thrombin and the effective competition of aptamer binding to immobilized thrombin by hirudin (Paborsky et al., 1993) indicated that the fibrinogen-binding exosite is important for aptamer binding.

5.4.3 Alanine Scanning Mutagenesis

To characterize the aptamer binding site in more detail, a rational mutagenesis strategy was designed to systematically scan the surface of human α-thrombin to identify residues important for inhibition by the thrombin aptamer (Tsiang et al., 1995). Residues were selected for mutagenesis on the basis of charge (Arg, Lys, Asp, Glu, His), polarity (Ser, Thr, Gln, Asn, Tyr, Trp) and solvent exposure, as these residues are likely to be important for the binding of charged ligands and less problematic with respect to non-specific structural perturbations (Bowie et al., 1990). Thrombin's charged and polar residues were ranked according to their fractional accessibility and the 70 residues with a fractional accessibility of >35% were selected for mutation. Arg36a, located at the junction between the A and B chains of a thrombin was not mutated because it is likely to be required for the processing of prethrombin to the mature α-thrombin. However, several residues (Arg20, Lys65, Arg68, Lys77) in the fibrinogen binding exosite implicated in binding the thrombin aptamer from binding competition, chemical modification protection (Liu and Gilbert, 1994) and site-directed mutagenesis (Wu et al., 1992) were added for mutation. The solution of the crystal structure of a complex of human α-thrombin and the thrombin aptamer indicated that the thrombin aptamer may interact with two sites on α-thrombin, the fibrinogen-binding exosite and the heparin-binding site (Kaillathe et al., 1993). Three residues not already included for mutation (His66, Arg98, Trp249) but predicted to make electrostatic interactions with the thrombin aptamer were also mutated. Thus a total of 76 residues were divided into 41 mutants with 1, 2, or 3 residues mutated simultaneously when clustered together.

All the mutants were capable of cleaving S-2238, thus indicating relative preservation of tertiary structure. The ability of the thrombin aptamer 15-mer to inhibit the mutant thrombins was assessed in a fibrinogen clotting assay. Three mutants had almost no clotting activity and were not tested. Five mutants were inhibited by less than 30% in this assay, corresponding to mutations K65A, H66A, R70A, Y71A and R73A (Figure 5.6).

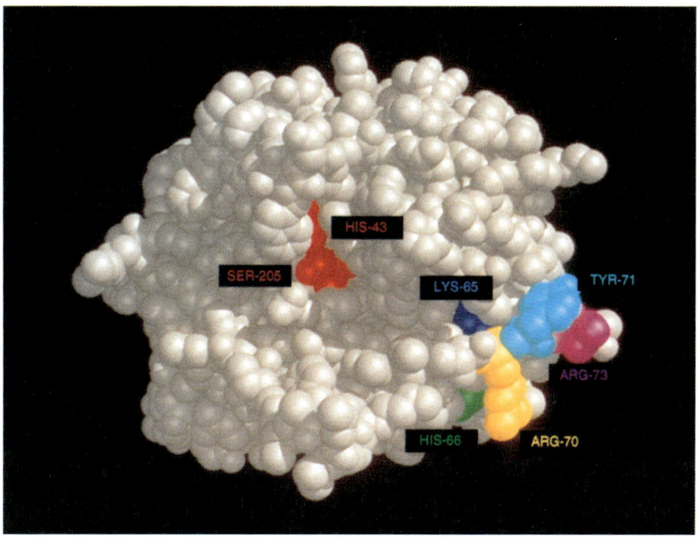

Figure 5.6: Model of thrombin showing the residues, K65, H66, R70, Y71 and R73, indicated by alanine scanning mutagenesis to be important for aptamer binding.

These five mutants were further assayed for their dose response to thrombin aptamer inhibition of clotting. Mutant R70A had a 2800-fold increase in IC_{50} followed by K65A, R73A, H66A and Y71A which increased the IC_{50} over wild type by 220-, 94-, 28- and 7-fold respectively. These five residues are localized to one site on thrombin in the fibrinogen-binding exosite. No residues in the putative heparin-binding exosite site were indicated as important for thrombin aptamer inhibition of thrombin. This finding is consistent with only one aptamer binding site on thrombin, however, the possibility of a second binding site not involved in thrombin inhibition cannot be ruled out by these experiments.

Only one of the residues identified by the lysine protection assay, Lys65, was identified in the alanine scanning mutagenesis. The other residue, Lys21, was not identified as important for thrombin aptamer inhibition. This result suggests that while Lys21 is protected from chemical modification in the presence of the aptamer, it is not directly or significantly involved in binding.

5.5 Determination of K_d and K_i

5.5.1 Surface Plasmon Resonance

Thrombin-aptamer interactions were characterized using surface plasmon resonance techniques (Davis, 1994). By using bimolecular interaction analysis (BIA) to study protein-aptamer interactions it was possible to calculate the kinetic and apparent affinity constants for the binding of thombin aptamer 15-mer to thrombin. The interactions were monitored by attaching single-stranded biotinylated DNA aptamer to a sensor chip via streptavidin and injecting thrombin across this surface. The interaction was also monitored by immobilizing thrombin on the sensor surface via PPACK, an irreversible active site inhibitor, and evaluating aptamer binding (Figure 5.7).

The dissociation rate constants (1.8×10^{-3}/s) were similar for both experiments. The association rates (2.9×10^5/Ms and 1.30×10^6/Ms respectively) were dependent on the surface. Thus the dissociation constant (K_d) determined for the thrombin aptamer interaction was 6.2nM using the biotinylated aptamer and 1.4nM using PPACK immobilized thrombin as the surface. These results are in good agreement with competition equilibrium binding experiments that showed binding of ^{125}I-labeled diisopropyl phosphoryl-thrombin to thrombomodulin was completely inhibited by the thrombin aptamer 15-mer with a K_d value of 2.7nM (Wu et al., 1992).

In order to more directly address the question of thrombin aptamer binding at the heparin-binding exosite as observed in the crystal structure (Kaillathe et al., 1993), biotinylated aptamer was immobilized on a sensor surface via streptavidin

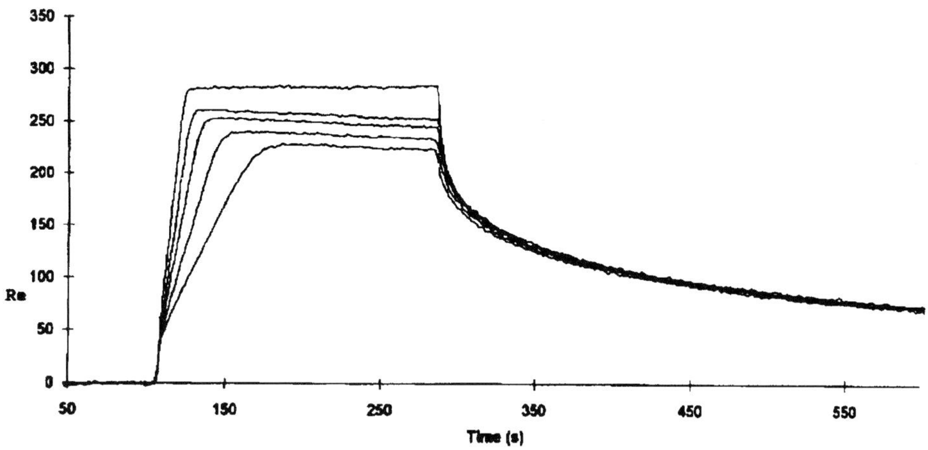

Figure 5.7: Sensorgram of thrombin-aptamer interaction by surface plasmon resonance. Aptamer (100–500 nM) injected over PPACK/thrombin surface.

and tested for binding of various thrombin mutants (Gibbs, personal communication). Data from the alanine scanning mutagenesis experiment showed that the thrombin aptamer has significantly decreased affinity for thrombin mutated in the fibrinogen-binding exosite (R70A, R73A, and R70E) while the active site and presumably the heparin-binding site of thrombin are largely unaffected. Thus, the use of these thrombin mutants should allow detection of aptamer binding at the heparin-binding exosite without the complication of aptamer binding at the fibrinogen binding exosite. However, at 250 nM R70A, only 1% of the binding observed with wild type thrombin was obtained, indicating that no significant aptamer binding outside the fibrinogen-binding exosite existed. This result was consistent with all of our solution phase data and suggests that the binding of the thrombin aptamer at the heparin-binding exosite is most likely an artifact of crystallization.

5.5.2 Platelet Thrombin Receptor Peptide Assay

Most thrombin inhibitors are active site inhibitors and their inhibition constants (K_i) can be determined by measuring the inhibitors effect on the cleavage rates of small chromogenic substrates. Because the thrombin aptamer does not inhibit by binding in the active site or altering the active site significantly, a different assay is required to determine the K_i. The thrombin clotting time assay, which we had previously used to demonstrate aptamer inhibition of thrombin, is not appropriate for K_i determinations because it does not directly measure fibrinogen cleavage products but rather detects the end product of the multi-step process of fibrin strand formation. Therefore, we developed a new HPLC-based assay that measures the rate of cleavage of the recently identified thrombin substrate, the platelet thrombin receptor peptide (PTRP) (Vu et al., 1991). The extracellular amino-terminal extension of the platelet thrombin receptor contains the thrombin cleavage site LDPR/S which is critical for receptor activation, and a sequence that is similar to the carboxyl tail of the leech anticoagulant hirudin and functions by binding to thrombin's fibrinogen-binding exosite. Consistent with the thrombin aptamer binding to the fibrinogen binding exosite, the thrombin aptamer inhibits thrombin induced platelet aggregation (Wu et al., 1992; Griffin et al., 1993). Thus the overlapping binding sites of the PTRP and the thrombin aptamer allow the use of the PTRP as a substrate in a thrombin aptamer inhibition assay. From the amino-terminal sequence of the receptor we synthesized a 21-mer peptide ATLDPR/SFLLRNPNDKYEPFW. The intact 21-mer and the 15-mer and 6-mer cleavage products can be easily separated and quantitated by reversed-phase HPLC. The K_m of the 21-mer peptide for thrombin was determined to be 20 ± 3 μM. Using this assay in a standard Lineweaver-Burk analysis we determined a K_i of 1.4 ± 0.1 nM for the 15-mer which was the same as the Kd determined by surface plasmon resonance for the thrombin aptamer and PPACK immobilized thrombin. As expected, because of their overlapping binding sites, the data also confirmed the thrombin aptamer to be a competitive inhibitor of PTRP cleavage.

5.6 *In vitro* Activity

5.6.1 Inhibition of Clot-Bound Thrombin

Heparin, the most commonly used antithrombin agent, is well known to be ineffective in treating arterial thrombosis (Yasuda et al., 1988; Topol et al., 1989). This ineffectiveness is likely due to the relative resistance of thrombin bound to fibrin clots or the exposed subendothelial matrix to heparin-antithrombin III (ATIII) inhibition (Weitz et al., 1990; Bar-Shavit et al., 1989; Hanson and Harker, 1988). In addition, heparin's anticoagulant function is neutralized by platelet factor 4 (PF4) released from activated platelets (Béguin et al., 1989). Thus, the incidence of bleeding and thrombocytopenia in some patients associated with the use of heparin (Warkentin and Kelton, 1989) and ineffectiveness in arterial thrombosis has led to a significant effort in the development of new antithrombin and antiplatelet agents.

The inhibition of clot-bound thrombin by the thrombin aptamer was compared with heparin and the synthetic thrombin inhibitor D-phenylalanyl-L-prolyl-L-arginyl chloromethyl ketone (PPACK) in an *in vitro* clot-bound thrombin assay measuring fibrinopeptide A (FPA) generation (Li et al., 1994). Heparin, which was used at clinically relevant concentrations of 0.2U/mL and 0.4U/mL, prolonged the activated partial thromboplastin time (aPTT) to ~2.0 and 3.4 times baseline, respectively. At this concentration, heparin completely inhibited the activity of free thrombin (0.05nM) added to plasma, which generated similar amounts of FPA as in the clot-bound thrombin system. However, heparin at 0.2U/mL was ineffective in inhibiting clot-bound thrombin (6.5% inhibition), and at 0.4 U/mL, it had only a modest inhibitory effect (34% inhibition). In contrast, the thrombin aptamer and PPACK significantly inhibited clot-bound thrombin (79.7% inhibition and 95% inhibition respectively). The thrombin aptamer and PPACK were used at a concentration (3 mM) that had approximately equivalent anticoagulant effect *in vitro* when compared with heparin prolonging the aPTT to 2.8 times baseline for both PPACK and the thrombin aptamer. The data indicate that clot-bound thrombin is protected form inhibition by heparin antithrombin III (ATIII) at clinically therapeutic heparin concentrations, whereas it is susceptible to inhibition by the thrombin aptamer, suggesting that the thrombin aptamer may be a superior anti-thrombotic agent to heparin.

5.6.2 Reduction of Arterial Platelet Thrombus Formation

To extend these *in vitro* observations to a clinically more relevant model, the thrombin aptamer was studied in an *ex vivo* whole artery angioplasty perfusion

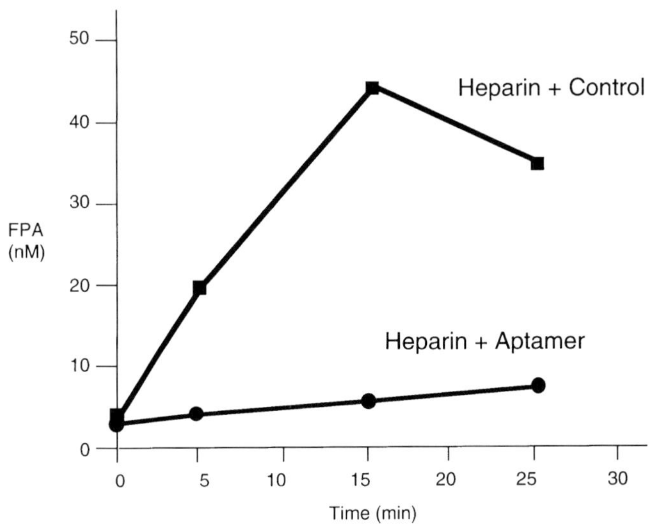

Figure 5.8: Effect of the thrombin aptamer on FPA generation in an *ex vivo* whole-artery angioplasty model. At the times indicated, aliquots of plasma were removed and amount of FPA assayed (mean ± SEM = 4).

model using radiolabeled platelets (Li et al., 1994). In this model, heparinized human whole blood was perfused over a balloon-angioplasty-injured rabbit aorta. Even in the presence of high-dose heparin (2U/mL), a significant amount of FPA was generated (Figure 5.8), suggesting that fibrin- or matrix-bound thrombin was resistant to heparin-ATIII. Similar to the clot bound thrombin data, the thrombin aptamer markedly suppressed the generation of FPA.

We also measured the efficacy of the thrombin aptamer in preventing platelet thrombus formation over the injured arterial segments. Compared to a scrambled DNA control, the thrombin aptamer (2μM) reduced platelet deposition by 34.5 ± 5% at low (200/sec) shear rates. At higher shear rates (850/sec), which more closely approximate those encountered in stenosed coronary arteries, the inhibition was more apparent at 61.6 ± 10.5%. The extent of inhibition by the thrombin aptamer was similar to that previously observed with PPACK and suggests that the thrombin aptamer can effectively inhibit thrombin and reduce platelet deposition at sites of vascular injury. In this model system the ssDNA thrombin aptamer is a potent inhibitor of clot- and matrix-bound heparin-resistant thrombin and effectively reduced arterial platelet thrombus formation.

5.7 *In vivo* Activity

5.7.1 Pharmacokinetic Studies

The ssDNA thrombin aptamer represents a novel class of thrombin inhibitor and it is important to study its *in vivo* anticoagulant properties and pharmacokinetic parameters (Griffin et al., 1993). To study the *in vivo* time course of thrombin aptamer-mediated anticoagulation we infused the thrombin aptamer into four cynomolgus monkeys at a rate of 0.3mg/kg/min over a period of one hour. Saline alone was also infused into one animal as a control. Anticoagulation was rapidly achieved with a plateau reached within ten minutes at ~2.0 times control prothrombin time (PT). After the infusion was stopped, there was a rapid reversal of the anticoagulant effect such that at ten minutes after the infusion, the PT had returned to baseline (Figure 5.9).

The pharmacokinetic data indicated that the *in vivo* half-life of the thrombin aptamer was extremely short. Based on an exponential fit of the data, the half-life was determined to be 108 ± 14sec. A variable rate of infusion gave a linear dose response relationship between the thrombin aptamer infusion rate and prolongation of plasma PT. All the animals tolerated the infusion of the thrombin aptamer without any problems and remained well after the experiments.

Blood samples from the variable dose infusion experiment were also used to study the effect of the thrombin aptamer on thrombin- and collagen-induced

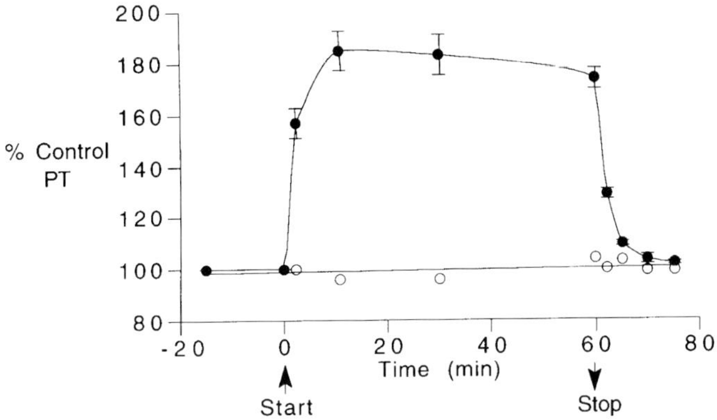

Figure 5.9: Time course of thrombin aptamer mediated anticoagulation (PT) during intravenous infusion (aptamer 0.3 mg/kg/min, closed circle, mean ± SD, n = 4; saline, open circle). Start and Stop indicate the beginning and end of the thrombin aptamer infusion time.

platelet aggregation. The thrombin aptamer effectively inhibited thrombin-induced platelet aggregation while even at the highest dose of thrombin aptamer, platelets responded normally to collagen, indicating that the thrombin aptamer inhibitory effect is specific for thrombin interaction with the platelet thrombin receptor, and not due to nonspecific inhibitory effects on platelets. Similar to the anticoagulant effect, platelets responded normally to thrombin 10 minutes after the infusion was stopped, again illustrating the short *in vivo* half-life of the thrombin aptamer.

Previous studies have shown that unmodified ssDNA oligonucleotides have a short half-life of <1 min following intravenous bolus injection in mice or rats (Emlen and Mannik, 1991; de Smidt et al., 1991). Our results are consistent with those observations. HPLC analysis that directly measures the plasma concentration of the thrombin aptamer showed that the short half-life directly correlated with decreasing plasma concentration (Cundy et al., manuscript in preparation). The short half-life does not appear to be due to rapid degradation of the ssDNA aptamer by plasma exonuclease activity (Shaw et al., 1995). In contrast to RNA, ssDNA is much more resistant to exonucleases and the thrombin aptamer 15-mer has an *in vitro* half-life in monkey plasma of >30 minutes and in human plasma >60 minutes. Furthermore, the infusion of a 3′-phosphorothioate thrombin aptamer, stable to exonuclease degradation, did not show a markedly increased *in vivo* half-life. Several studies suggest that cellular uptake of unmodified phosphodiester-linked oligonucleotides involves endocytosis, apparently mediated by specific saturable cell surface receptors (Loke et al., 1989; Yukabov et al., 1989; Shoji et al., 1992). Fast uptake of ssDNA by the liver has been observed after intravenous bolus injection in mice (Emlen and Mannik, 1991). Also wide tissue distribution of modified oligonucleotides after intravenous injection has been reported (de Smidt et al., 1991; Chen et al., 1990; Zendegui et al., 1992; Agrawal et al., 1991). Further studies addressing the clearance of the thrombin aptamer are described in section 5.7.3.

5.7.2 Regional Anticoagulation

Hemodialysis in the presence of systemic anticoagulation with heparin carries a significant risk of clinically important hemorrhage (10% to 19%) (Zusman et al., 1981). Regional anticoagulation is desirable in clinical situations in which systemic anticoagulation poses a substantial risk for major hemorrhage, such as patients with uremic pericarditis, acute renal failure, active gastrointestinal bleeding, and recent cardiovascular surgery. To utilize the short *in vivo* half life of the thrombin aptamer, its ability to achieve regional anticoagulation in an extracorporeal circuit was tested (Griffin et al., 1993). Hemofiltration was performed in the sheep with the thrombin aptamer being infused just proximal to the hemofiltration unit. Clotting times were determined from samples taken simultaneously from a peripheral vein and the extracorporeal circuit. During the course of hemofiltration, the blood circulating in the extracorporeal circuit was effectively anticoagulated (PT 40–45 sec, baseline PT 21.7 sec); however, because of the short *in vivo* half-life, the aptamer did not accumulate in the body and the blood in the animal was not significantly anticoagulated (Figure 5.10).

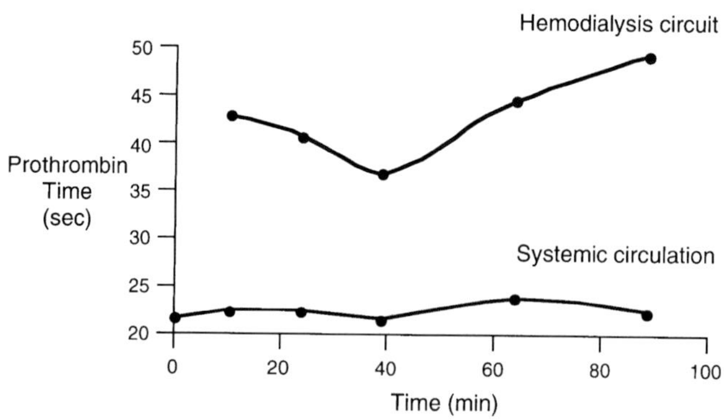

Figure 5.10: Regional anticoagulation during hemofiltration in the sheep. Blood samples for PT analysis were drawn from the circuit just prior to the hemofilter and simultaneously from the femoral venous catheter to measure the extent of systemic anticoagulation.

Thus, the systemic PT remained essentially unchanged clearly demonstrating regional anticoagulation in an extracorporeal circuit by the thrombin aptamer. In contrast, the use of heparin results in anticoagulation of the systemic circulation in addition to the circuit. Regional anticoagulation would greatly minimize the risk of bleeding during dialysis and be advantageous to some patients with kidney failure who cannot tolerate heparin because of bleeding complications.

5.7.3 Cardiopulmonary Bypass

Patients undergoing cardiopulmonary bypass (CPB) require systemic anticoagulation. Heparin has traditionally been used for this purpose, but is contraindicated in patients with a history of heparin-induced thrombocytopenia and/or thrombosis; heparin in such patients may lead to severe bleeding or arterial or venous thrombosis. In a small number of patients heparin therapy can also be ineffective due to a deficiency of antithrombin III, the plasma cofactor required

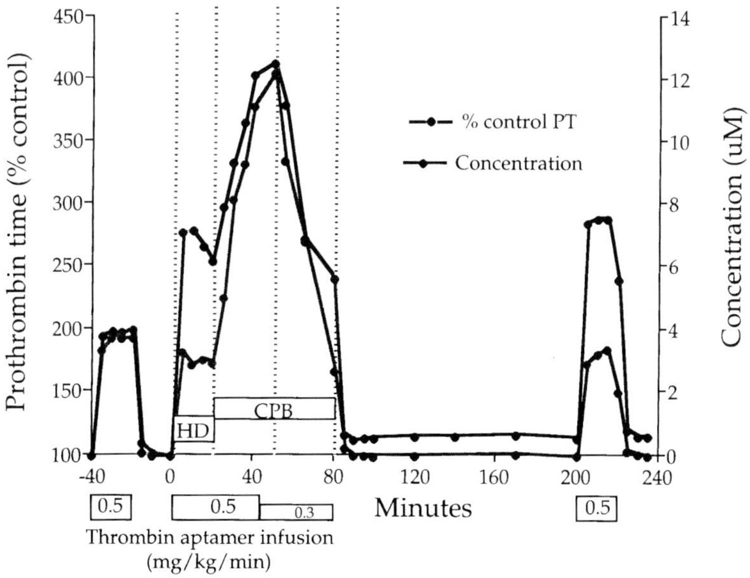

Figure 5.11: Anticoagulant profile of the thrombin aptamer in a single animal. The vertical axis on the left is the degree of anticoagulation as assessed by prothrombin time, with the corresponding plasma concentration of the aptamer shown on the vertical axis on the right. HD = hemodiluted state (*i.e.*, pump prime circulated, partial CPB); CPB = total cardiopulmonary bypass. The horizontal bars on the bottom of the panel indicate the aptamer infusion rate in mg/kg/min.

for the anticoagulant effect of heparin. Also, the use of heparin during CPB requires protamine sulfate to reverse its anticoagulation effect; not only is protamine associated with hypotension due to systemic vasodilation and myocardial depression, but it may also rarely be associated with a catastrophic anaphylactic reaction (Frater et al., 1984). A short acting thrombin inhibitor, such as the thrombin aptamer, may potentially serve as a substitute for heparin in this and other clinical situations and obviate the need for reversal of anticoagulant effect at the end of bypass.

We tested the thombin aptamer in a canine CPB pilot study to determine its anticoagulant efficacy, resultant changes in coagulation parameters, and the aptamer's clearance mechanisms and pharmacokinetics (Figure 5.11) (DeAnda et al., 1994). Initially seven dogs were studied: Four received varied doses of the aptamer (to establish the pharmacokinetic profile), and three received heparin. Four other dogs then underwent CPB receiving a constant infusion of the aptamer pre-CPB (to characterize baseline coagulation status), with partial CPB and hemodilution, during 60 min of total CPB, and finally, after a 2 hour recovery period.

At a 0.5mg/kg/min dose, the % control prothrombin time (PT) rose with aptamer infusion to ~200% and then increased further to ~280% with infusion of the pump prime into the systemic circulation and resultant dilution of the coagulation factors (hemodilution). This dilution was also observed in the slight decrease in aptamer concentration as determined by HPLC. The PT was even more prolonged during total CPB to >400%This later increase in PT paralleled a rise in HPLC-determined plasma concentration of the thrombin aptamer during total CPB. The calculated plasma elimination half-life increased from 1.9min during baseline conditions to 7.7min during total bypass, suggesting that the pulmonary microcirculation plays a role in aptamer clearance. Importantly, the coagulation profile returned to normal levels within 5 minutes of ending the aptamer infusion. There was no excessive intraoperative blood loss and postoperative bleeding (chest tube output) was minimal (45–75mL/hr), and serum chemistry profiles were normal. These preliminary studies indicate that the thrombin aptamer is both safe and effective as an anticoagulant in this short-term canine CPB model and has predictable pharmacokinetics. Furthermore, the significant increase in the half-life during total bypass suggests that the pulmonary circulation plays a major role in clearance of the aptamer.

5.8 Conclusions

Using aptamer technology we were able to isolate from a large random pool of ssDNA a subset of molecules with increased affinity and specificity for thrombin, a protein not known to interact physiologically with nucleic acid. This activity is retained in a 15-mer GGTTGGTGTGGTTGG (thrombin aptamer) which folds into a novel G-quartet structure. Although the thrombin aptamer selection was based solely on the ability to bind thrombin, the thrombin aptamer is also functionally active as a potent anticoagulant *in vitro* and *in vivo* that in some settings may be superior to existing anticoagulants. Thus, the use of aptamer technology targeted against thrombin has provided a promising and unique antithrombin lead.

The thrombin aptamer provides a lead for further drug design by a variety of chemical modifications, including the phosphodiester backbone, the bases, and the sugar moieties, to enhance its pharmacologic and therapeutic properties. Modifications that introduce hydrophobic groups and replace specific phosphates in the backbone without disrupting G-quartet formation have been found to increase potency *in vitro* and *in vivo.*

Acknowledgements

The authors would like to acknowledge all those at Gilead who have contributed to the thrombin aptamer project. This work was supported in part by SBIR Grant No. 1R43 HL48431–01 and 2R44 HL48431–02.

References

Agrawal, S., Temsamani, J. and Tang, J.Y. (1991) Pharmacokinetics, biodistribution, and stability of oligodeoxynucleotide phosphorothioates in mice, Proc. Natl. Acad. Sci. USA *88*, 7595–7599.

Bar-Shavit, R., Eldor, A. and Vlodavsky, I. (1989) Binding of thrombin to subendothelial extracellular matrix. Protection and expression of functional properties, J. Clin. Invest. *84*, 1096–1104.

Béguin, S., Lindhout, T. and Hemker, H.C. (1989) The effect of trace amounts of tissue factor on thrombin generation in platelet rich plasma, its inhibition by heparin, Thromb. Haemost. *61*, 25–29.

Bischofberger, N. and Shea, R.G. (1992) Oligonucleotide-based therapeutics, In: Nucleic Acid Targeted Drug Design (Propst, C.L. and Perun, T.J.; eds.) pp. 579–612, Marcel Dekker, New York, NY.

Bock, L., personal communication.

Bock, L.C., Griffin, L.C., Latham, J.A., Vermaas, E.H. and Toole, J.J. (1992) Selection of single-stranded DNA molecules that bind and inhibit human thrombin, Nature 355, 564–566.

Bowie, J.U., Reidhaar-Olson, J.F., Lim, W.A. and Sauer, R.T. (1990) Deciphering the message in protein sequences: Tolerance to amino acid substitutions, Science *247*, 1306–1310.

Chang, J.-Y. (1989) The hiridin binding site of human α-thrombin, J. Biol. Chem. *264*, 7141–7146.

Chang, T.-L., Feinman, R.D., Landis, B.H. and Fenton, J.W., II. (1979) Antithrombin reactions with alpha and gamma-thrombin, Biochem. *18,* 113–119.

Chen, T.L., Miller, P.S., Ts'o, P. and Colvin O.M. (1990) Disposition and metabolism of oligodeoxynucleoside methylphosphonate following a single iv injection in mice, Drug Metabolism and Disposition *18,* 815–818.

Church, F.C., Pratt, C.W., Noyes, C.M., Kalayanamit, T., Sherrill, G.B., Tobin, R.B. and Meade, J.B. (1989) Structural and functional properties of human α-thrombin, phosphopyridoxylated α-thrombin and γ_T-thrombin, J. Biol. Chem. *264,* 18419–18425.

Cundy, K.C., Shaw, J.P., Fishback, J.A., Bock, L., Griffin, L.C. and Lee, W.A., manuscript in preparation. Davis, S. (1994) Kinetic characterization of thrombin-aptamer interactions, J. of Biomolecular Interaction Analysis. Application note #*305,* 29.

de Smidt, P.C., Doan, T.L., de Falco, S. and Van Berkel, T.J.C. (1991) Association of antisense oligonucleotides with lipoproteins prolongs the plasma half-life and modifies the tissue distribution, Nucl. Acid. Res. *19,* 4695–4700.

DeAnda Jr., A., Coutre, S.E., Moon, M.R., Vial, C.M., Griffin, L.C., Law, V.S., Komeda, M., Leung, L.L.K. and Miller D.C. (1994) Pilot study of the efficacy of a thrombin inhibitor for anticoagulation during cardiopulmonary bypass, Annals of Thoracic Surgery *58,* 344–350.

Ellington, A.D. and Szostak, J.W. (1990) In vitro selection of RNA molecules that bind specific ligands, Nature *346,* 1104–1110.

Emlen, W. and Mannik, M. (1991) Effect of DNA size and strandedness on the in vivo clearance and organ localization of DNA, Clin. Exp. Immunol. *56,* 185–192.

Farley, R.A., Tran, C.M., Carilli, C.T., Hawke, D. and Shively, J.E. (1984) The amino acid sequence of a fluorescein-labeled peptide from the active site of (Na,K)-ATPase, J. Biol. Chem. *259,* 9532–9535.

Fenton, J.W., II, Olsen, T.A., Zabinski, M.P. and Wilner, G.D. (1988) Anion-binding exosite of human alpha-thrombin and fibrinogen recognition, Biochem. *27,* 7106–7112.

Fenton, J.W., II. (1981) Thrombin specificity, Ann. N.Y. Acad. Sci. *370,* 468–495.

Frater, R.W.M., Oka, Y., Hong, Y., Tsubo, T., Loubger, P.G. and Masone, R. (1984) Protamine induced circulatory changes, J. Thoracic Cardiovascular Surgery *87,* 687–92.

Furie, B. and Furie, B.C. (1988) The molecular basis of blood coagulation, Cell *53,* 505–518.

Gailani, D. and Broze, G.J. (1991) Factor XI activation in a revised model of blood coagulation, Science *253,* 909–912.

Gibbs, C., personal communication.

Griffin, L.C., Tidmarsh, G.F., Bock, L.C., Toole, J.J. and Leung, L.L.K. (1993) In vivo anticoagulant properties of a novel nucleotide-based thrombin inhibitor and demonstration of regional anticoagulation in extracorporeal circuits, Blood *81,* 3271–3276.

Grutter, M.G., Priestle, J.P., Rahuel, J., Grossenbacher, H., Bode, W., Hofsteenge, J. and Stone, S.R. (1990) Crystal structure of the thrombin-hirudin complex: A novel mode of serine protease inhibition, Embo J. *9,* 2361–2365.

Guschlbauer, W., Chantot, J.-F. and Thiele, D. (1990) Four stranded nucleic acid structures 25 years later; from guanosine gels to telomere DNA, J. Biomolec. Struc. Dynamics *8,* 491–511.

Hanson, S.R. and Harker, L.A. (1988) Interruption of acute platelet-dependent thrombosis by the antithrombin D-phenylalanyl L-prolyl-L-arginyl chloromethyl ketone, Proc. Natl. Acad. Sci. USA *85,* 3184–3188.

Hardin, C.C., Henderson, E., Watson, T. and Prosser, J.K. (1991) Monovalent cation induced structural transitions in telomeric DNA's G-DNA folding intermediates, Biochemistry *31,* 4460–4472.

Henderson, E.R., Hardin, C.C., Walk, S.K., Tinoco Jr., I. and Blackburn, E.H. (1987) Telomeric DNA oligonucleotides form novel intramolecular structures containing guanine guanine basepairs, Cell 51, 899–908.

Hultman, T., Stahl, L., Moks, T. and Uhlen, M. (1988) Approaches to solid phase sequencing, Nucleosides and Nucleotides *7,* 629–638.

Kaillathe, P., Kanagapushpam, P.P., Ferrara, J.P., Sadler, J.E. and Tulinsky, A. (1993) The structure of a thrombin inhibited by a 15-mer single-stranded DNA aptamer, J. Biol. Chem. *268,* 17651–17654.

Krawczyk, S.H., Bischofberger, N., Griffin, L.C., Law, V.S., Shea, R.G. and Swaminathan, S. (1995) Structure-activity study of oligodeoxynucleotides which inhibit thrombin, Nucleosides & Nucleotides, *14,* 1109–1116.

Li, W.X., Kaplan, A.V., Grant, G.W., Toole, J.J. and Leung, L.L.K. (1994) A novel nucleotide based thrombin inhibitor inhibits clot-bound thrombin and reduces arterial platelet thrombus formation, Blood *83,* 677–682.Liu, Z. and Gilbert, W. (1994) The yeast KEM1 gene encodes a nuclease specific for G4 tetraplex DNA: Implication of *in vivo* functions for this novel DNA structure, Cell *77,* 1083–1092.

Loke, S.L., Stein, C.A., Zhang, X.H., Mori, K., Nakanishi, M., Subasinghe, C., Cohen, JS. and Neckers, L.M. (1989) Characterization of oligonucleotide transport into living cells, Proc. Natl. Acad. Sci. USA *86,* 3474–3478.

Macaya, R.F., Schultze, P., Smith, F.W., Roe, J.A. and Feigon J. (1993) Thrombin-binding DNA aptamer forms a unimolecular quadruplex structure in solution, Proc. Nat., Acad. Sci. USA *90,* 3745–3749.

Mann, K.G., Nesheim, M.E., Church, W.R., Haley, P. and Krishnaswamy, S. (1990) Surface-dependent reactions of the vitamin K-dependent complexes, Blood *76,* 1–16.

Messmore, H.L., Fareed, J., Zabinski, M.P., Orfei, P., Kniffin, J. and Fenton, J.W., II. (1979) Quantification of antithrombin III with various molecular forms of human thrombins, Fed. Proc. *38,* 758.

Mitchinson, C., Wilderspin, A.F., Trinnaman, B.J. and Green, N.M. (1982) Identification of a labelled peptide after stoichiometric reaction of fluorescein isothiocyanate with the Ca^{2+} dependent adenosine triphosphatase of sarcoplasmic reticulum, FEBS Lett. *146,* 87–92.

Paborsky, L., McCurdy, S., Griffin, L.C., Toole, J.J. and Leung, L.L.K. (1993) The single-stranded DNA aptamer-binding site of human thrombin, J. Biol. Chem. *268,* 20808–20811.

Phillips, N.F.B. (1988) The ATP/AMP binding site of pyruvate, phosphate dikinase: Selective modification with fluorescein isothiocyanate, Biochemistry *27,* 3314–3320.

Rogers, S.J., Pratt, C.W., Whinna, H.C. and Church, F.C. (1992) Role of thrombin exosites in inhibition by heparin cofactor II, J. Biol. Chem. *267,* 3613–3617.

Shaw., J.P., Fishback, J.A. Cundy, K.C. and Lee, W.A. (1995) In vitro metabolic stability in plasma and serum of a novel oligonucleotide inhibitor of thrombin, Pharmaceutical Research in press.

Sheehan, J.P., Wu, Q. and Sadler, J.E. (1991) Abstract in Blood 78, Suppl. *1,* 277a.

Shoji, Y., Akhtar, S., Periasamy, A., Herman, B. and Juliano, R.L. (1992) Mechanism of cellular uptake of modified oligodeoxynucleotides containing methylphosphonate linkages, Nucleic Acid Res. *19,* 5543–5550.

Shuman, M.A. (1986) Thrombin cellular interactions, Ann. N.Y. Acad. Sci. *485,* 228–239.

Topol, E.J., George, B.S., Kereiakes, D.J., Stump, D.C., Candela, R.J., Abbotsmith, C.W., Aronson, L., et al. (1989) A randomized controlled trial of intravenous plasminogen activator and early intravenous heparin in acute myocardial infarction, Circulation *79,* 281–286.

Tsiang, M., Jain, A.K., Dunn, K.E., Rojas, M.E., Leung, L.L.K., and Gibbs, C.S. (1995) Functional Mapping of the Surface Residues of Human Thrombin, J. Biol. Chem. *270,* in press.

Vu, T.-K., Wheaton, V.I., Hung, D.T., Charo, I. and Coughlin, S.R. (1991) Domains specifying thrombin-receptor interaction, Nature *353,* 674–677.

Wang, K.Y., Krawczyk, S.H., Bischofberger, N., Swaminathan, S. and Bolton, P.H. (1993a) The tertiary structure of a DNA aptamer which binds to and inhibits thrombin determines activity, Biochemistry *32,* 11285–11292.

Wang, K.Y., McCurdy, S.N., Shea, R.G., Swaminathan, S. and Bolton, P.H. (1993b) A DNA aptamer which binds to and inhibits thrombin exhibits a new structural motif for amino acid substitutions dissociate fibrinogen-clotting and thrombomodulin-binding activities of human thrombin, Proc. Natl. Acad. Sci.USA *88*, 6775–6779.

Wu, Q., Tsiang, M. and Sadler, J.E. (1992) Localization of the single-stranded DNA binding site in the thrombin anion-binding exosite, J. Biol. Chem. *267*, 24408–24412.

Yasuda, T., Gold, H.K., Fallon, J.T., Leinbach, R.C., Guerrero, J.L., Scudder, L.E., Kanke, M., Shealy, D., Ross, M.J., Collen, D. and Coller, B.S. (1988) Monoclonal antibody against the platelet glycoprotein (GP) IIb/IIIa receptor prevents coronary artery reocclusion after reperfusion with recombinant tissue-type plasminogen activator in dogs, J. Clin. Invest. *81*, 1284–1291.

Yukabov, L.A., Deeva, E.A., Zarytova, F., Ivanova, E.M., Ryte, A.S., Yurchenko, L.V. and Vlassov, V.V. (1989) Mechanism of oligonucleotide uptake by cells: Involvement of specific receptors? Proc. Natl. Acad. Sci. USA *86*, 6454–6458.

Zahler, A.M., Williamson, J.R., Cech, T.R. and Prescott, D.M. (1991) Inhibition of telomerase by G-quartet DNA structures, Nature *350*, 718–720.

Zendegui, J.G., Vasquez, K.M., Tinsley, J.H., Kessler, D.J. and Hogan, M.E. (1992) In vivo stability and kinetics of absorption and disposition of 3 phosphopropyl amine oligonucleotides, Nucleic Acid Res. *20*, 307–314.

Zusman, R.M., Rubin, R.H. Cato, A.E., Cocchetto, D.M., Crow, J.W. and Tolkoff-Rubin, N. (1981) Hemodialysis using prostacyclin instead of heparin as the sole antithrombotic agent, N. Engl. J. Med. *304*, 934–939.

C Phage Display of Peptide Libraries

6 Structural and Functional Constraints in the Display of Peptides on Filamentous Phage Capsids

Gianni Cesareni, Olga Minenkova, Luciana Dente, Gioacchino Iannolo, Adriana Zucconi, Manuela Helmer Citterich, Alessandra Lanfrancotti , Luisa Castagnoli and Costantino Vetriani

6.1 Introduction

Recently a series of different approaches grouped under the general term of "*in vitro* evolution" have raised considerable interest (Kauffman, 1993; Bartel and Szostak, 1993; Tuerk and Gold, 1990; Smith and Scott, 1993; Lowman et al., 1991; Stemmer, 1994). In synthesis, the idea is to mimic the natural selection process, in the test tube, in order to either select molecules with new properties or acquire information about the possible routes explored by nature in evolving contemporary structures, functions and organisms.

This approach is extremely powerful, inasmuch as evolutionary events that took billions of years to occurr can be reproduced in a very short time span. Several genetic systems have been explored in implementing this approach and, although they differ widely, depending on the specific question asked and on the selection applied, they share the common characteritics of 1) being easily and extensively modifiable, and 2) having a short duplication time (Bartel and Szostak, 1993; Tuerk and Gold, 1990; Smith and Scott, 1993; Maruyama et al., 1994). This permits several selection cycles to be applied on vast collections of variants of the specific *organism* utilized in just a few days.

When the property searched is the ability to bind (recognize) a given target molecule, phages make attractive candidates for exploitation as scaffolds in the construction of large collections of protein structures, each with different binding properties. In addition to their structural simplicity and short replication cycle, the main advantage that phages offer is the possibility to link, in a simple way, the selection to the amplification step in the *in vitro* evolution protocol.

Although phage lambda has also been explored as carrier for protein or protein domains (Maruyama et al., 1994), filamentous phage is by far the best organism for the applications that take advantage of the ability to construct large molecular repertoires by modifying the solvent exposed part of a phage envelope (Smith and Scott, 1993). This is because, uniquely among the most studied bacteriophages, spectroscopic techniques have permitted to draw a molecular model of the filamentous capsid, thus providing a structural base for rational modifications (Banner et al., 1981; Marvin et al., 1994; Glucksman et al., 1992; McDonnell et al., 1993).

Furthermore, filamentous phage are unusually resistant to chemical or heat treatment. As a consequence,the function under study can, if necessary, be exposed to harsh treatment (for instance to release the bound phage) without interfering with phage viability. On the other hand, the assembly process of filamentous phage requires traslocation of the capsid protein across the bacterial membrane, without damaging the integrity of the cell envelope. As a consequence, some peptides may not be efficiently displayed on filamentous phage capsid, either because they can not be translocated across the cytoplasmic membrane or because they are sensitive to degradation by periplasmic proteases. In these cases different phage display systems, like the lambda one, might turn out to be valuable.

Several peptide libraries have already been constructed by fusing random peptides to the amino terminus of pIII or close to the amino terminus of pVIII (Scott and Smith, 1990; Cwirla et al., 1990; Devlin et al., 1990; Felici et al., 1991). These libraries are successfully being searched with a variety of target molecules. The full exploitation of this technology, however, requires a deeper understanding of the phage structure and life cycle. Here we will briefly review what is known about filamentous phage biology and structure and we will report the work done in our laboratory aimed at exploring the limits in our ability to modify the phage capsid without interfering with phage production and infectivity.

6.2 The Phage Life Cycle

Filamentous bacteriophage are a class of ssDNA phage that can only infect male bacteria, either F^+ or Hfr (for reviews about phage biology and morphogenesis see Rasched and Oberer, 1986; Russel, 1991; Model and Russel, 1988). Under the electron microscope, the appearance of an M13 bacteriophage is that of a flexible filament 6–10 nm thick and 900 nm long.

Immediately after a phage particle has infected its host, the single stranded phage chromosome is made double stranded by host enzymes and then, with the

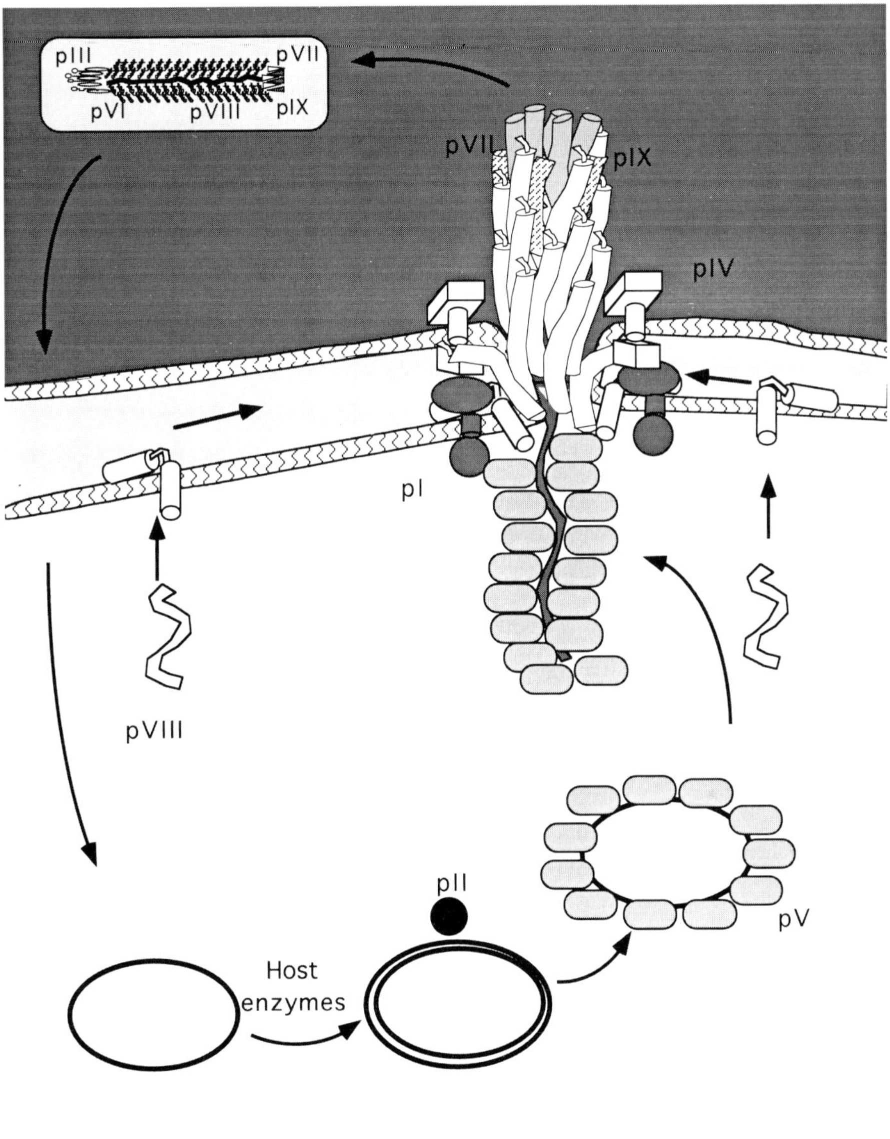

Figure 6.1: Filamentous phage life cycle. Infection occurs via the F pilus that is not illustrated in this cartoon. The details of the assembly process, as illustrated here are mostly speculative. The experimental evidence is cited in several reviews (16, 17, 18).

involvement of the phage encoded protein pII, replicated several times to build a pool of double stranded replicative form (RF) DNA (Figure 6.1). In the meantime the remaining phage genes are transcribed and translated. The phage products that eventually will contribute to form the viral particle (pVIII, pIII, pVII, pIX and pVI) accumulate in the inner membrane. Later in the infection cycle, the product of gene V (pV) determines the switch from double stranded to single stranded DNA replication. The newly synthesized ssDNA is complexed with pV and finally phage assembly takes place at specialized sites in the membrane where newly synthesized coat proteins and other proteins necessary for phage assembly and extrusion are stored (Lopez and Webster, 1983; Lopez and Webster 1985). Although the details of the assembly process are not fully understood, it is established that two non-coat proteins, the product of gene IV and gene I, cooperate with bacterial proteins to produce phage particles.

Both pI and pIV are trans-membrane proteins: pI is localized in the inner membrane with a 100 amino acid cytoplasmic tail (Guy-Caffey et al., 1992) while pIV co-purifies with the outer membrane (Brissette et al., 1990). Genetic evidence indicates there is a functional interaction between the amino-terminal domain of pI and the carboxy-terminal domain of pIV in the periplasmic space (Russel, 1993). pIV is homologous to proteins that facilitate the export of secreted proteins in gram negative bacteria and forms a homo-multimeric complex of 10 to 12 subunits which may work as a gated hydrophilic channel through which the assembling virion reaches the extra-cellular environment (Kazmierczak et al., 1994; Russel, 1994).

The comprehension of this phage extrusion mechanism should help us in rationalizing the limits that we encounter when modifying phage capsid and in designing better vectors for phage display.

From the point of view of phage display technology, an important characteristic of the filamentous phage life cycle is that infection does not lead to bacterial lysis or cell death. The growth of a phage clone can be visualized as a plaque because infected cells have a longer doubling time, and are characterized by a less turbid area on a confluent lawn of non-infected bacteria. Furthermore, if the phage chromosome carries a drug resistance marker, infected cells can be selected and propagated on antibiotic plates in the same way as any plasmid transformed cell.

The phage receptor on the bacterial surface is the tip of a thread-like structure, the sex pilus encoded by the F episome in male strains. At the beginning of the infective process, receptor recognition triggers the retraction of the pilus, the stripping of the major coat protein from the DNA and its injection into the cytoplasm, by a mechanism which ,at present, is poorly characterized.

6.3 A Low Resolution Model

Most of the 20 Å thick cylinder that wraps the DNA is built from approximately 2700 copies of a 50 amino acid long peptide: the product of gene VIII, the major coat protein. The two tips of the virion are capped by two molecular complexes formed by the association between the products of gene VI and gene III at the proximal end, that leaves the cell last, and the product of genes VII and IX at the distal end.

No high resolution structure of the virion is available. However, a combination of low resolution data from a variety of physical techniques, X-ray diffraction of fibers (Banner et al., 1981; Glucksman et al., 1992), solid state NMR (McDonnell et al., 1993) and Raman spectroscopy (Aubrey and Thomas, 1991) has permitted a molecular model of the virion to be proposed and refined. The radius of the outside surface of the cylinder is approximately 33 A, while the length of the cylinder depends on that of the packaged DNA since approximately 0,435 wild type coat proteins are required to neutralize each phosphate on the DNA backbone. The coat proteins that make up a large fraction of the mass of the virion are arranged in a helical symmetry with a five fold rotational axis. They overlap each other and form a pattern that is reminiscent of the scales of a fish. Each pVIII peptide is a single gently curved alpha helix running from Pro6 to the carboxyl end.

Although the fiber diffraction experiments are consistent with the amino-terminal residues being also alpha helical, solid state NMR indicates that residues 1 to 5 are mobile. The major coat protein can be schematically divided into four parts: a mobile acidic amino-terminus of five residues; a relatively polar alpha helix from Pro5 to Tyr24; a hydrophobic helix extending from Ala25 to Ala35 and an amphipatic carboxy-terminal helix, with positive groups positioned to neutralize the phosphates in the nucleic acid from residue Thr36 to Ser50.

No direct piece of evidence is available to help draw the structures of the minor coat proteins. In the model illustrated in Figure 1, we assume that Makowski's proposal, based on some amino acid sequence homology between the minor proteins pVII , pIX and pVI and the major coat protein is correct and as a consequence we draw the pVII and pIX complex as a hydrophobic molecular cork made of five helices arranged with a five fold symmetry (Makowski, 1992). The orientation of the peptides is similar to that of pVIII with the amino-terminus outside.

6.4 Extending the Amino-Terminus of pVIII

Early attempts (Ilichev et al., 1989; Greenwood et al., 1991; Felici et al., 1991) to utilize the pVIII backbone to display combinations of amino acid side chains concentrated on the possibility of modifying the solvent exposed amino-terminus of the protein. Although the details of the fusion technology and the exact insertion sites differed in the various experiments, it was soon realized that most peptides longer than six amino acids were not tolerated when one attempted to display them all over the surface of the filament. This problem was overcome by engineering hybrid phages where a few copies of the amino terminally extended pVIII protein were interspersed in an otherwise wild type phage coat (Felici et al., 1991).

This was achieved by exploiting a phagemid system where the recombinant hybrid pVIII protein is synthesized by a gene on a phagemid, while the wild type protein is provided by a helper phage. By utilizing this system we have shown that more than 70% of a collection of random nonapeptides can be efficiently displayed by fusion to the major coat protein. Large libraries of random peptides of this sort have been constructed and successfully used.

However the density of the displayed peptide can vary dramatically from few copies per virion up to approximately 30% of the total coat protein depending on the specific amino acid sequence. Furthermore, it is often difficult to obtain reproducible peptide densities in equivalent preparations of phage displaying the same peptide. Although this does not represent a major problem it would be desirable, for certain applications, to be able to produce phage that reproducibly display any peptide at the same high density.

The limitations encountered in engineering phages displaying peptides longer than six amino acids at the amino terminus of pVIII do not depend on the selected insertion site, since different insertion schemes have proven equally unsuccessful (Ilichev et al., 1989; Greenwood et al., 1991; Felici et al., 1991). In an attempt to identify the most tolerant sites we have recently explored three different strategies to display heptapeptides of random sequence by fusion to three different regions of the amino terminus of pVIII (Figure 6.2) (Iannolo et al., 1994, and our unpublished results). The first experiment consisted of inserting an oligonucleotide designed to encode seven random amino acids in place of the dipeptide Ala7Lys8 of the wild type sequence of pVIII in M13. Only 5% of the clones obtained produced infective particles. Significantly, most of the viable phage displayed a heptapeptide that reintroduced, at the carboxyl end either AlaLys or AlaArg (Figure 6.2c). This result proves that the chemical properties of the side chains of the two residues that form the amino end of the pVIII helix have some important functions.

In a second attempt we explored a different region by moving the insertion site three residues toward the amino terminus. In this case the percentage of

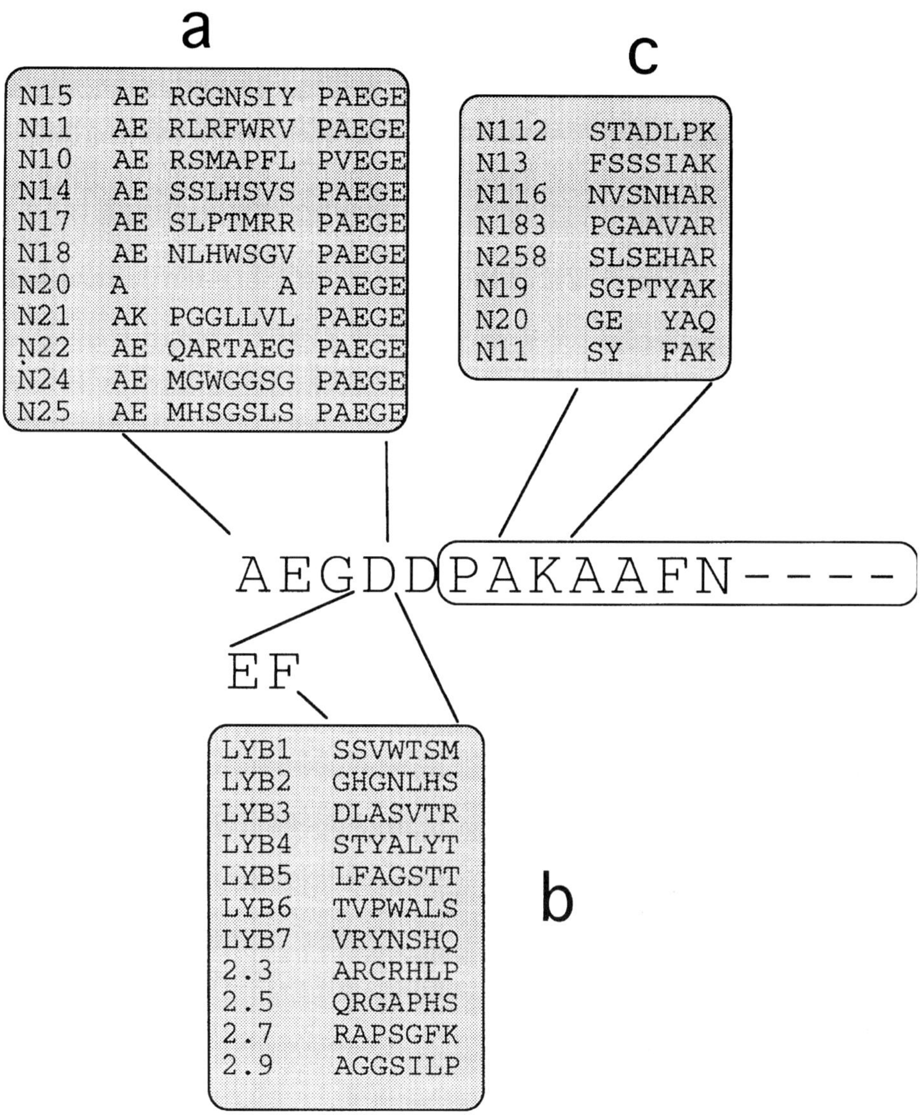

Figure 6.2: Insertion of heptapeptides into different regions of the amino terminus of the major coat protein. In large letters are the 12 amino terminal residues of pVIII. The residues in α-helix conformation are boxed. The amino acid sequences of the peptide inserts in randomly picked clones of three different peptide libraries are framed by the shaded boxes labelled a, b and c.

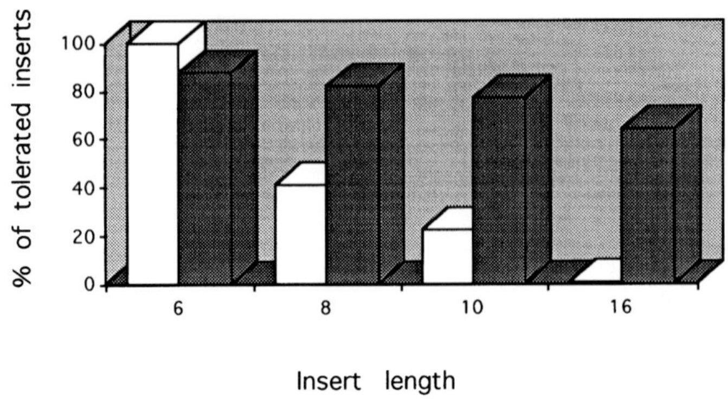

Figure 6.3: Percentage of tolerated random inserts in the amino terminus of pVIII as a function of insert length. In each experiment 100 clones were tested and the fraction that was able to produce phage particles determined by measuring their ability to transduce tetracycline resistance. The shaded bars represents the expected fraction of inserts containing non sense codons.

clones that proved able to produce viable phage increased to 40%, and analysis of the peptide sequence displayed by the viable clones did not reveal any striking sequence pattern (Figure 6.2b).

Finally, we designed a third strategy that would allow us to construct phage where the inserted peptide is fused exactly to the amino terminus of the mature coat protein without interrupting its amino acid sequence. Also in this case approximately 30% of the engineered clones were able to produce infective particles (Figure 6.2a). We therefore conclude that, by choosing the appropriate insertion site, it is possible to construct large libraries of phage displaying all over the phage capsid many copies of, at least, heptapeptides with minor limitations.

The major limiting factor tingo display peptides by fusion to the major coat protein is the length of the added peptides. Hexapeptides are practically always tolerated irrespective of their sequence, while the percentage of peptides of eight, ten and sixteen amino acids whose insertion still allow formation of infective particles, drops to 40, 20 and 1 respectively (Figure 6.3). Apart from the absence of unpaired Cys, no other prohibited patterns or regularities could be deduced from the amino acid sequence of the peptides displayed by the viable phages, indicating that the specific amino acid sequence plays only a fine tuning role, probably by modulating the conformation of the peptide itself or its interaction with the phage capsid (Iannolo et al., 1994).

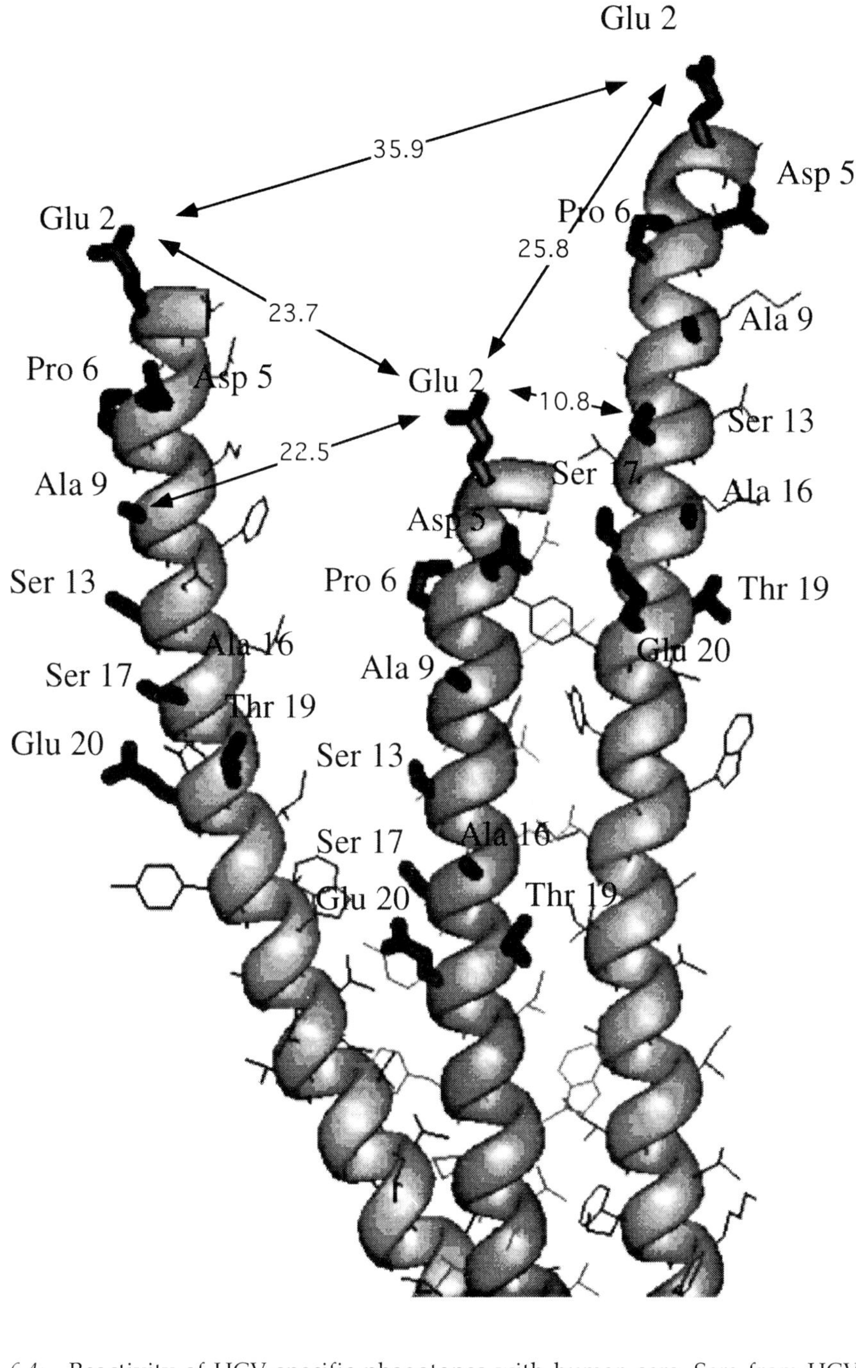

Figure 6.4: Reactivity of HCV-specific phagotopes with human sera. Sera from HCV-infected patients (top panel) and healthy individuals (bottom panel) was measured by ELISA. Positive reactions to the HCV-specific phagotoes (listed on the left of each panel) are represented by black squares; negative reactions are indicated with grey squares.

6.5 Modifying the Surface of Filamentous Phage by Amino Acid Substitution

Most attempts to build phage collections displaying molecular repertoires have considered filamentous phage as "passive" carriers of peptides or protein domains. In a first approximation it is assumed that the structural environment in which the displayed peptide is embedded, the capsid surface, does not influence its conformation and ability to bind a target. It is apparent however (see for instance Figure 6.4) that the binding of any phage displayed peptide to a given molecule, especially in the case of pVIII display, is affected by the chemical characteristics of the phage surface that either directly interact with the protein target or indirectly modulate the peptide properties by influencing its conformation. It follows that it should in principle be possible to alter the affinity of a peptide for its target by changing the characteristics of the capsid residues that surround it. One could immagine having a collection of phages each characterized by one or several substitutions in the solvent exposed region of the major capsid protein from whuch to select "the best" structural context to enhance the properties of a given peptide.

To this end we have explored by random mutagenesis the solvent exposed region of the pVIII protein (from Ala1 to Tyr 24) to assess which side chains could be modified without affecting the ability to produce infective particles.

The results are illustrated in Figure 6.5. The amino-terminal five residues can be extensively modified (Boraschi et al., 1989) without affecting the ability of the coat protein to assemble into an infective particle. However, substitutions that increase the global charge of this pentapeptide by 2, from –3 to –1, almost invariably fail to support phage assembly, probably because of a defect in the transport of the precursor of the major coat protein across the cytoplasmic membrane. Moving down and entering the sequence of the amphipatic α-helix, we infer from the insertion experiments previously described, although we have not proved this by single amino acid substitutions, that Ala7 plays an important role as the positive charge of Lys8. Most of the remaining residues up to Tyr21 can also be changed. Although still incomplete and insufficient to draw statistically meaningful conclusions, our results are in general agreement with the model proposed by Marvin and colleagues (Marvin et al., 1994) with the residues that are proposed to be exposed to the solvent tolerating more numerous and less conservative substitutions. On the contrary, residues that, according to the model, are involved in intra-molecular contacts in the capsid structure are less tolerant. Not surprisingly Pro is not accepted in the α-helix. Taken together these results and those of the insertion experiments suggest that it should soon be possible, by taking advantage of the phage model, to add some rational engineering to the successful random approach in the construction of more versatile libraries and in the maturation of the affinity of the displayed peptides.

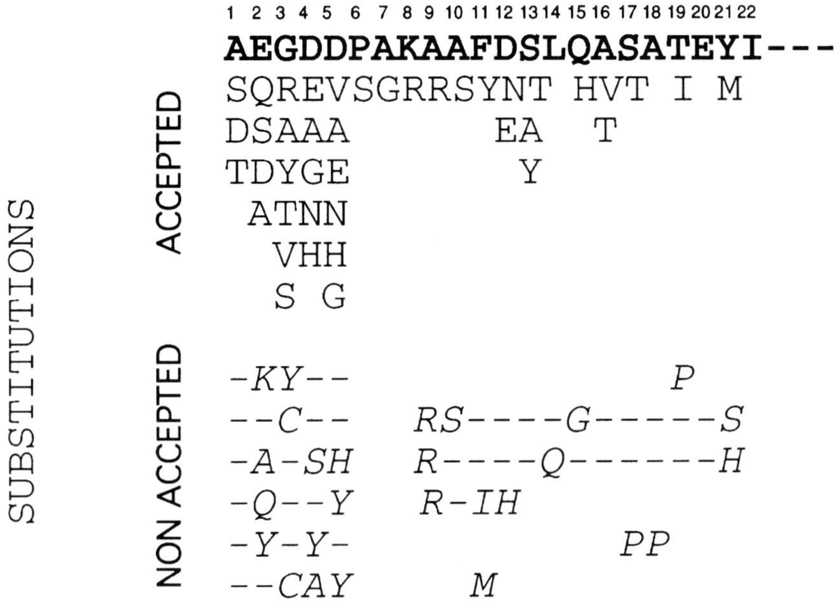

Figure 6.5: Amino acid substitutions in the solvent exposed surface of the major coat protein.The amino acid sequence of the 22 amino terminal residues of pVIII are printed in two different tones. In black are the residues that, according to the model of Marvin and colleagues (9) are involved in intersubunit contacts while in grey are the solvent exposed residues. The accepted amino acids, indicated below each wild type residue, were found either as single amino acid substitutions or in combination with other substitutions. The botton block of sequences represents combination of residues that resulted in clones unable produce infective particles.

6.6 Can the Remaining Minor Proteins Support Modification?

Many different peptides have been successfully displayed by fusion either to pIII or to pVIII. These two capsid proteins with their proved tolerance to peptide insertions and with their different copy number range seem to cover, in a complementary fashion, all one would wish from a display system.

Is there any need to improve our display technology by exploring new insertions sites in any of the three remaining minor coat proteins? We believe that with the development of the applications and scope of display technology it will become more and more desirable to engineer multivalent structures.

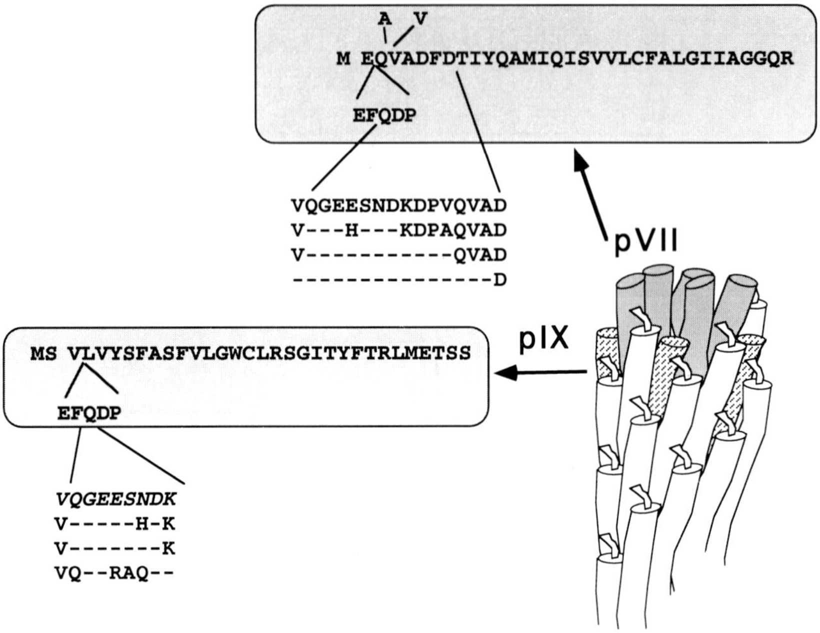

Figure 6.6: Peptide insertions in pVII and pIX. On the left is a cartoon of the structure of the distal end of a filamentous phage essentially as proposed by Makowski (27). Underneath the amino acid sequence of pVII and pIX we have indicated the amino acid sequence of the inserts that do not affect the ability of the phage to produce infective particles. In italics is the amino acid sequence of the peptide, from interleukin 1β that we have unsuccesfully tried to engineer in the indicated region of pIX.

To this end we have recently explored the feasibility of displaying peptides by fusion to pVII and pIX. Figure 6.6 illustrates a model of the structure of the distal end of the phage filament essentially as proposed by Makowski (Makowski, 1992). The helical arrangement of the major coat protein is terminated by a complex formed by five copies of the product of gene VII and gene IX. By analogy with pVIII, both proteins are assumed to be helical with the amino-terminus pointing outside and possibly exposed to the environment.

In the absence of any information about the sensitivity of these proteins to amino acid change, our first experiment was rather bold. since we attempted to insert a pentapeptide immediately after the amino-terminal Met in the case of pVII and after Met Ser in the case pf pIX. The sequence of the pentapeptide was dictated by our intent to engineer in the coding sequence an amber codon flanked by a *BamH*1 and an *EcoR*1 site to facilitate further engineering.

Both mutant proteins assembled into phage structures and were able to form plaques.

We thus conclude that the amino-terminus of these two minor proteins can be extensively modified. Although this result does not prove the proposed topology, with an exposed amino terminus, it is in better agreement with this model than with the alternative one, since it would seem improbable that five extra residues are forced into the already crowded inside of the phage. The insertion of the pentapeptide into pVII is not completely "painless" for the phage, since in two independent constructions we have observed the accumulation of a second spontaneous mutation that changes the negatively charged Glu2 into a hydrophobic side chain.

To discover whether the insertions at the amino-terminus of these two proteins would result in the the inserted peptide being presented to the environment in a form available for interaction with other soluble molecules, we have attempted to engineer a nonapeptide that is recognised by a monoclonal antibody raised against the Il1β molecule (Bora-schi et al., 1989) in the insertion sites previously described.

Most of the clones that have been characterized, however, carried a different deletion of the Il1b peptide and of the flanking coat protein sequence. We have to conclude that either because of its length or sequence this peptide is not tolerated by the phage when is fused to the amino terminus of either pVII or pIX (Figure 6.6). Furthermore we could not show, by biopanning experiments that the single clone that we found carrying the correct insert at the amino-terminus of pVII could be enriched by biopanning-type experiments.

Nonetheless the clones carrying rearranged pVII or pIX hybrid peptides indicate that at least one nonapeptide is tolerated by pIX, while the longest functional pVII that we have seen is 13 amino acids longer than the native protein. These extensively deleted clones also demonstrate that residues 3 to 8 of pVII are not essential for function.

References

Aubrey, K.L. and Thomas, G.J. J. (1991) Raman spectroscopy of filamentous bacteriophage Ff (fd, M13, f1) incorporating specifically-deuterated alanine and tryptophan side chains. Assignments and structural interpretation, Biophys J *60*, 1337–1349.

Banner, D.W., Nave, C. and Marvin, D.A. (1981) Structure of the protein and DNA in fd filamentous bacteriophages, Nature *289*, 814–816.

Bartel, D.P. and Szostak, J.W. (1993) Isolation of new rybozimes from a large pool of random sequences, Science *261*, 1411–1418.

Boraschi, D., Volpini, D., Villa, L., Nencioni, L., Scapigliati, G., Nucci, D., Antoni, G., Matteucci, G., Cioli, F. and Tagliabue, A. (1989) A monoclonal antibody to the IL-1β peptide 163–171 blocks adjuvancy of IL-1β in vivo, J. Immunol. *143*, 131–134.

Brissette, J.L. and Russel, M. (1990) Secretion and membrane integration of a filamentous phage encoded morphogenetic protein, J. Mol. Biol. *211*, 565–580.

Cwirla, S.E., Peters, E.A., Barret, R.W. and Dower, W.J. (1990) Peptides on phage: a vast library of peptides for identifying ligands, Proc. Natl. Acad. Sci. USA *87*, 6378–6382.

Devlin, J.J., Panganiban, L.C. and Devlin, P.E. (1990) Random peptide libraries: a source of specific protein binding molecules, Science *249*, 404–406.

Felici, F., Castagnoli, L., Musacchio, A., Jappelli, R. and Cesareni, G. (1991) Selection of antibody ligands from a large library of oligopeptides expressed on a multivalent exposition vector, J. Mol. Biol. *222*, 301–310.
Glucksman, M.J., Bhattacharjee, S. and Makowski, L. (1992) Three dimensional structure of a cloning vector. X-ray diffraction studies of filamentous bacteriophage M13 at 7 Å resolution, J. Mol. Biol. *226*, 455–470.
Greenwood, J., Willis, A.E. and Perham, R.N. (1991) Multiple display of foreign peptides on a filamentous bacteriophage. Peptides from Plasmodium falciparum circumsporozoite protein as antigens, J. Mol. Biol *220*, 821–827.
Guy-Caffey, J.K., Rapoza, M.P., Jolley, K.A. and Webster, R.E. (1992) Membrane localization and topology of a viral assembly protein, J. Bacteriol *174*, 2460–2465.
Iannolo, G., Minenkova, O., Petruzzelli, R. and Cesareni, G. (1994) Modifying filamentous phage capsid: Limits in the size of the major capsid protein, I. Mol. Biol. (1995) *248*, 835–849
Ilichev, A.A., Minenkova, O.O., Tat'kov, S.L., Karpyshev, N.N., Eroshkin, A.M., Petrenko, V.A. and Sandakhchiev, L.S. (1989) Production of a viable variant of the M13 phage with a foreign peptide inserted into the basic coat protein, Dokl. Akad. Nauk. S.S.S.R. *307*, 481–483.
Kauffman, S.A. (1993) Applied molecular evolution, J. Theor. Biol *157*, 1–7.
Kazmierczak, B.I., Mielke, D.L., Russel, M. and Model, P. (1994) pIV, a filamentous phage protein that mediates phage export across the bacterial cell envelope, forms a multimer, J. Mol. Biol. *238*, 187–198.
Lopez, J. and Webster, R. (1985) Assembly site of bacteriophage f1 corresponds to adhesion zones between the inner and outer membrane of the host cell, J. Bacteriol. *163*, 1270–1274.
Lopez, J. and Webster, R.E. (1983) Morphogenesis of filamentous bacteriophage f1: orientation of extrusion and production of polyphage, Virology *127*, 177–193.
Lowman, H.B., Bass, S.H., Simpson, N. and Wells, J.A. (1991) Selecting high-affinity binding proteins by monovalent phage display, Biochem. *30*, 10832–10838.
Makowski, L. (1992) Terminating a macromolecular helix. Structural model for the minor proteins of bacteriophage M13, J. Mol. Biol. *228*, 885–892.
Maruyama, I.N., Maruyama, H.I. and Brenner, S. (1994) Lambda foo: a lambda vector for the expression of foreign proteins, Proc. Natl. Acad. Sci. USA *91*, 8273–8277.
Marvin, D.A., Hale, R.D., Nave, C. and Helmer Citterich, M. (1994) Molecular models and structural comparisons of native and mutant class I filamentous bacteriophages, J. Mol. Biol. *235*, 260–286.
McDonnell, P.A., Shon, K., Kim, Y. and Opella, S.J. (1993) fd coat protein structure in membrane environments, J. Mol. Biol. *223*, 447–463.
Model, P. and Russel, M. (1988) Filamentous bacteriophage, In: The bacteriophage (Calendar, R.; ed.), pp. 375–456
Rasched, I. and Oberer, E. (1986) Ff coliphages: structural and functional relationship, Microbiol. Rev. *50*, 401–427.
Russel, M. (1991) Filamentous phage assembly, Mol. Microbiol. *5*, 1607–1613.
Russel, M. (1993) Protein-protein interaction during filamentous phage assembly, J. Mol. Biol. *231*, 689–697.
Russel, M. (1994) Phage assembly: a paradigm for bacterial virulence factor export? Science *265*, 612–614.
Scott, J.K. and Smith, G.P. (1990) Searching for peptide ligands with an epitope library, Science *249*, 386–390.
Smith, G.P. and Scott, J.K. (1993) Libraries of peptides and proteins displayed on filamentous phage, Methods Enzymol. *217*, 228–257.
Stemmer, W.P.C. (1994) Rapid evolution of a protein in vitro by DNA shuffling, Nature *370*, 389–391.
Tuerk, C. and Gold, L. (1990) Systematic evolution of ligands by exponential enrichment: RNA ligands to bacteriophage T4 DNA polymerase, Science *249*, 505–510.

7 Conformationally Defined Peptide Libraries on Phage: Selectable Templates for the Design of Pharmacological Agents

Maurizio Sollazzo, Elisabetta Bianchi, Franco Felici, Riccardo Cortese, Antonello Pessi

7.1 Introduction

The extraordinary diversity of biological activities mediated by peptides is directly correlated to the vast number of possibilities that exist for their amino acid sequences and three-dimensional structures. As a result, knowing the active conformation of a given polypeptide is a fundamental advance towards understanding its biological activity. Although a wealth of biophysical techniques exist to study polypeptide conformation, a major obstacle to the characterization of small peptides is their intrinsic flexibility. Small peptide ligands in solution often are an ensemble of almost iso-energetic conformers, and the question arises as to which of these is biologically relevant (Marshall, 1992), an issue complicated by the fact that many biologically active peptides have multiple sites of interaction (Hruby et al., 1984). The binding of a peptide to its receptor may change the conformation of the ligand and possibly that of the receptor itself: as a result, peptide dynamics becomes an important factor (Schiller and DiMaio, 1983). At present, the fundamental issue of what precisely is the relationship between peptide primary structure, conformation and topography is still unsolved (Hruby et al., 1990).

7.2 From Peptides to Peptidomimetics

Despite these difficulties, peptides are one of the most studied compounds in medicinal chemistry and remain the richest and most ubiquitous source of lead compounds, and more so since biotechnology has become the driving force behind the selection and identification of novel peptide ligands (Olson et al., 1993). This potential has been further enhanced by the advent of phage display (Smith, 1991) and combinatorial synthetic peptide libraries (for a review see Gallop et al., 1994). Peptides, however, display very unfavourable pharmacological properties, like poor bioavailability, short duration of action and lack of oral activity. It is not surprising that increasing efforts are underway to rationally design and synthesize biologically active non-peptide analogs of peptides, now currently referred to as peptidomimetics (Hruby et al., 1990; Olson et al., 1993). Constructing the link between peptides and classical pharmaceuticals, as well as understanding how to accomplish all the steps involved, would allow the prolific number of lead molecules discovered to be converted into drug candidates.

The peptide-peptidomimetic transition can be divided into three steps: *(a)* identifying the amino acid side-chains responsible for activity ("pharmacophoric groups"); *(b)* establishing the spatial relationship between these groups ("pharmacophore model"); *(c)* peptidomimetic design, i.e. selecting templates for pharmacophoric groups which are suitable for reproducing the geometry of the pharmacophore model (Moore, 1994). The most difficult, and usually rate-limiting step is the second one, since only rarely can the biologically relevant peptide topography be deduced from direct observation of the receptor-ligand complex. Through synthetic organic chemistry, peptide chemists have learnt how to constrain the conformational freedom of polypeptides by using a wide variety of experimental strategies (Rizo and Gierash, 1992). Structure-activity relationship studies on these constrained analogs ultimately lead to deriving the active conformation and consequently the pharmacophore topography.

During the past decade the number of biochemical and biophysical studies based on constrained peptides has increased exponentially, and the variety of applications of these studies with them. Model constrained peptides have been used to study in detail the conformational preferences of natural and modified amino acid residues, affording new insight into the mechanisms of protein folding and molecular recognition (Rizo and Gierash, 1992). Active peptides have been designed and synthesized with specific structural features and used to define the conformational requirements for binding to corresponding receptor(s), leading to the development of potent agonists and antagonists (Beeley, 1994). The development of these and new applications will doubtless continue to grow, with the main focus on the design and synthesis of structurally defined peptides, as the last step in peptidomimetic design, i.e. the selection and synthesis of a peptide backbone-replacing organic template, is meeting with increasing success

(Kahn et al., 1988; Saragovi et al., 1991; Hirschmann et al., 1992a, b; Nolan et al., 1992; Smith et al., 1992; Sato et al., 1992; Ku et al., 1993; McDowell et al., 1994; Smith et al., 1994; Jameson et al., 1994).

7.3 Building Constraints in Phage-Displayed Polypeptides

The increased ability to identify peptide leads through phage display may be offset by the difficulties of converting peptide structural information to peptidomimetic molecules. If we wish to fully exploit the potential offered by genetics, the issue to be addressed is how to utilize a biological system consisting of naturally occurring amino acids only, for introducing constraints, without the tools offered by synthetic chemistry. For small peptide hormones, nature uses two alternatives to reduce the flexibility of polypeptide chains: the incorporation of cyclic amino acid residues (proline) and the formation of macrocycles through disulfide bonds. In these two ways, specific conformations of peptides and proteins are stabilized by reducing the entropy cost upon folding into such conformations. Attempts to introduce constraints in peptide libraries displayed on phage have indeed been made by introducing cysteine residues that flank the variable residues (O'Neil et al., 1992; Luzzago et al., 1993; McLafferty et al., 1993). Fusion phage containing a protein domain added to the pIII or the pVIII proteins that would act as "presentation scaffold"(Wetzel, 1991) are an alternative approach (Smith, 1991). The rationale behind this strategy is that although the number of possible conformations increases enormously with molecular size, proteins are much more prone than peptides to adopt well-defined structures. The basis for this behaviour is twofold. First, the primary driving force for adoption of a compact conformation in a protein, which is generally accepted to be the sequestration of the hydrophobic surface area, is less effective in directing peptide conformational preferences. Second, the intramolecular interactions that maintain a polypeptide chain in a specific folded structure are weak, and only a large number of such interactions can compensate for the conformational entropy lost when the polypeptide adopts a single major conformation. Some proteins, on the other hand, have been reported as being tolerant to the insertion of bioactive peptide sequences (Maeda et al., 1989; Freimuth et al., 1990; Sollazzo et al., 1990; Wolfson et al., 1991; Langen and Taylor, 1992; Roberts et al., 1992; Barbas et al., 1992, 1993; Lee et al., 1993; Zanetti et al., 1993).

Having more than one protein architecture amenable to structural determination and containing permissive sites for the display of libraries within a secondary structural motif could be of practical importance. We thus targeted two peptide topologies: the reverse turn and the α-helix. In the search for a suitable scaffolding for the display of constrained loops we drew inspiration from antibody variable domains (Pessi et al., 1993), and to build libraries of α-helical motifs we chose the zinc finger as a naturally occurring small protein domain (Bianchi et al., 1995).

7.4 The Minibody: An Engineered β-Pleated Scaffold for the Display of Reverse-Turn Motifs

The antibody variable domain, in simplified terms, can be considered as six randomized constrained loops mounted onto a rigid β-pleated framework. Antibodies, however, optimize conformational diversity, and appear capable of generating virtually any possible binding-site by combining the right amount of sequence randomization within a conserved structural framework with a reasonable degree of constraint to allow for loop movements, induced fit, and alternative packing geometry of V_H-V_L domains (Padlan, 1994). Reproducing such a spectacular ductility in a repertoire of smaller polypeptides would be an extraordinary feat. It has been shown that single variable domains may be sufficient to produce binding surfaces with reasonable affinities and specificities (Ward et al., 1989; Davie and Riechmann, 1994; Sun et al., 1994). These observations have gained further support by the discovery that camels can mount effective immune responses using antibodies consisting solely of heavy chains (Hamers et al., 1993; Muyldermans et al., 1994). In the attempt to make a β-pleated polypeptide that is smaller than a single domain, but still able to produce specific binding surfaces, we constructed the *minibody.* We wanted our molecule to be tolerant to sequence variability in restricted regions of the polypeptide, and to be able to predict the main chain conformation of the same regions based on the "canonical structures" model (Chothia and Lesk, 1987; Chothia et al., 1989; Tramontano et al., 1990). If successfully designed, such a β-sheet framework could serve as a platform for engineering complex binding surfaces using both rational and combinatorial strategies (Pessi et al., 1993). An additional consideration for choosing this particular folding motif was that synthetic strategies for designing peptidomimetics based on immunoglobulin-like reverse turns are object of intensive studies (for a review see: Taub and Greene, 1992; Saragovi et al., 1992). Because the de novo design of β-proteins is a very difficult matter (Richardson et al., 1992), we focused on the re-design of a known antibody variable domain as a structural template. We trusted that for this class of molecules, the essential requirements for the local stability of parts of the fold could be accurately identified (Tramontano et al., 1994). The outcome of our re-design effort was a small protein (61-res-

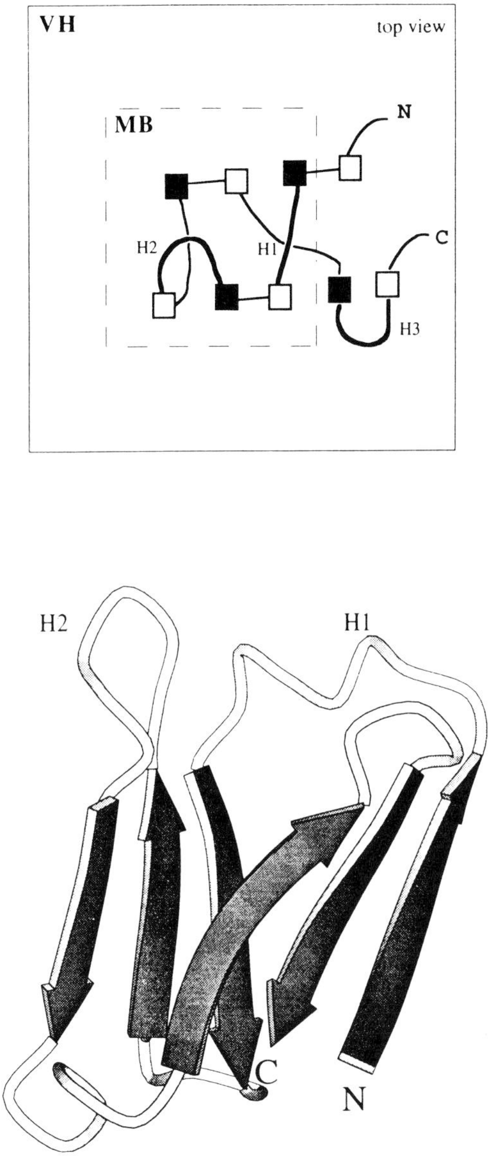

Figure 7.1: Schematic top view (adapted from Chothia and Lesk , 1987) of an antibody heavy chain variable domain (VH), H1, H2 and H3 correspond to the hypervariable regions. The dashed area marks the subdomain portion of V_H which corresponds to the minibody (MB). The gray boxes are the upcoming β-strands with respect to the plane of the figure whereas the clear boxes represent β-strands going in the opposite direction. Below is shown the structural model of the minibody (approximately perpendicular view), H1 and H2 loops, amino (N) and carboxy (C) termini are indicated. The picture was generated using MOLSCRIPT (Kraulis, 1991).

idues long), which consists of three β-strands from each of the two β-sheets of the variable domain of an antibody, along with the exposed hypervariable H1 and H2 regions, as shown in Figure 7.1.

We proved that both the metal-free and metal-bound forms of the minibody are compact globular monomers in solution. Far-UV circular dichroism (CD) measurements showed that the minibody has the expected amount of secondary β-structure, and denaturation studies with urea showed that the molecule unfolds with a co-operative transition: its $\Delta G_D^{H_2O}$ value, although lower than the values observed for natural proteins (5–15 kcal/mol), is very similar to that of the reduced immunoglobulin C_L domain (1.8 kcal/mol). The tertiary structural properties of the minibody were analyzed by near-UV CD spectroscopy. The asymmetry associated with the minibody's aromatic residues results in a negative CD band in the aromatic region of the spectrum, a clear indication of the existence of tertiary structure (Radford et al., 1992). These results were thus fully consistent with those expected for a compact, monomeric β-protein (Bianchi et al., 1994). Further evidence that the loop conformation is compatible with the predicted structure came from the successful engineering of a metal-binding site into the molecule through the introduction of three histidine residues at appropriate sites, one in H1 and two in H2, that display metal selectivities and the same order of relative affinities of those observed in carbonic anhydrase B, the template protein from which the metal-binding site was grafted (Pessi et al., 1993).

Unfortunately, the millimolar concentration indispensable for NMR spectroscopy studies was achieved only under very acidic conditions (pH2.8). The spectroscopic analysis carried out under these conditions showed the presence of some long-range NOEs and a number of slow exchanging amide protons, indicating the presence of a stable hydrophobic core even under these severe conditions, but the number of these long-range NOEs was not sufficient for detailed structural characterization (Barbato et al., unpublished observations). Although it is not known to what extent the results of this analysis may reflect the situation at neutral pH, we decided to use the NMR data to refine our minibody model and steer a second re-design cycle, this time aimed at stabilizing the molecule and improving solubility in more neutral pHconditions. We have recently produced a single amino acid mutant in the framework that shows a more co-operative denaturation transition, indicating that we succeeded in improving the hydrophobic packing of the minibody (Tramontano et al., in preparation).

Based on the above evidence, we concluded that design of the minibody was satisfactory. While waiting for a more stable NMR amenable variant, we reasoned that the parent molecule was good enough to go ahead with the idea of constructing a repertoire of minibody loop mutants, and to establish to what extent the minibody surface could be modulated to create affinity-selectable novel binding activities. To this end, a fusion phage was generated in which the

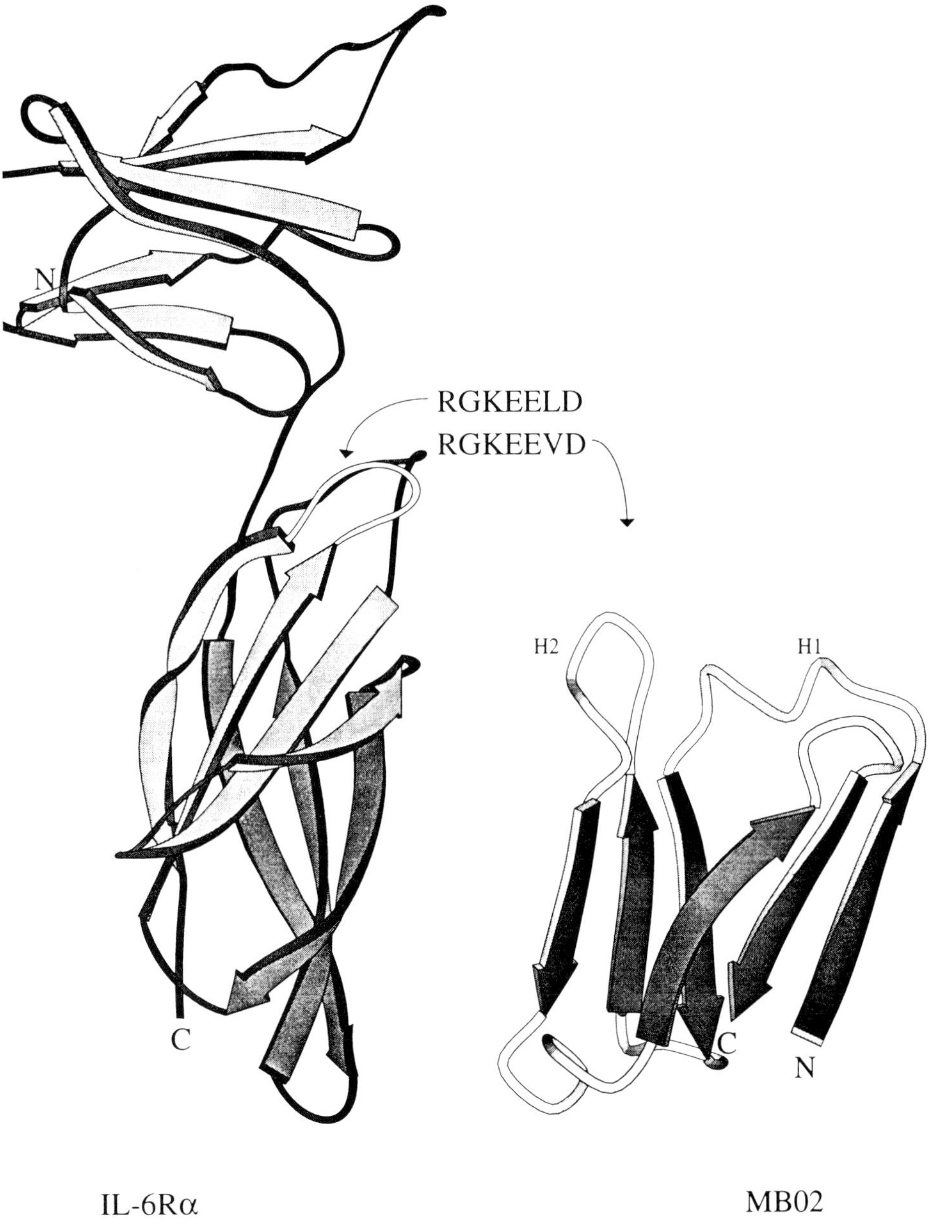

Figure 7.2: Receptor mimicry hypothesis. Structural models of the IL-6Rα (adapted from Savino et al., 1994) and of the affinity selected MB02. The amino acid sequences corresponding to H2 loop and the murine receptor region (residue 293–299) are indicated in single-letter code. The corresponding region of the human receptor has been shown to be involved in the binding to the cytokine (Yawata et al., 1993).

minibody was displayed onto the f1 phage surface as a hybrid with the pIII protein, and proven to be affinity-selectable using a specific monoclonal antibody (Venturini, et al., 1994).

As a first step toward an "all purpose" minibody repertoire, we generated a library of minibodies consisting of approximately 50 million members by randomizing the H1 and H2 loops (Martin et al., 1994). The question to be answered was whether this library was sufficiently complex to enable the affinity-selection of minibody-ligands with high affinity and specificity for a given target. To address this issue we used human Interleukin-6 (hIL-6) as a target molecule.

IL-6 is a multifunctional cytokine that plays a major role in the regulation of hematopoiesis, immune response, and acute phase reaction, involved in several pathological conditions (Akira et al., 1993). The biological activity of IL-6 is triggered by the sequential assembly of a receptor complex formed by a specific membrane-bound IL-6 receptor (IL-6Rα) and the signal transducing subunit gp130 (Kishimoto et al., 1994). Because of this mode of action, it has been proposed that specific inhibitors could be designed to neutralize IL-6 bioactivity by blocking the first event (Savino et al., 1994).

After multiple rounds of affinity selection using immobilized hIL-6 as a target, we isolated several clones with different properties. The best minibody molecule (MB02) bound specifically hIL-6 with a K_D value of 230nM, a remarkable result considering the limited size of the library. A particular characteristic of this clone was a striking sequence similarity between the MB02 H2 region and a stretch of amino acid residues of murine (m)IL6Rα. Because hIL-6 binds mIL-6Rα, and mutants of the corresponding region in the hIL-6Rα do not bind hIL-6 (Yawata et al., 1994), we reasoned that MB02 could be a functional receptor mimic (Figure 7.2). Further studies using CD spectroscopy proved that MB02 was folded with a secondary structure content similar to that of its parent molecule MB1. What is more important, its interaction with hIL-6 inhibits the cytokine from bind-ing to its cognate receptor with an IC_{50} of 2 μM *in vitro.* It was established furthermore, that the MB02 could block IL-6 dependent signal transduction at micromolar concentrations in tissue culture assay (Martin et al., 1994). It appears that MB02, by acting as a surrogate receptor, is capable of binding with the cytokine and effectively inhibiting hIL-6 biological activity.

These results show that from a collection of fifty million minibodies it was possible to select at least one ligand for hIL 6 with a dissociation constant value in the low micromolar range. This is to our knowledge the first example of the successful combination of rational design and a combinatorial approach to construct a small β-protein with a novel biological function. Because there is no *a priori* reason to believe that hIL-6 represents a unique example, we reasoned that the minibody library, at least in principle, could be used as a source of constrained polypeptide ligands for any desirable macromolecule.

7.5 The Zinc Finger: A Small Domain for the Display of Structurally Homogeneous α-Helical Motifs

In this case the chosen protein domain was the Cys_2His_2 'zinc-finger', one of the most common eukaryotic DNA-binding motifs (Pellegrino and Berg, 1991; Jacobs, 1992). The residues responsible for specificity, mainly located in the α–helical sub domain (Pavletich and Pabo, 1991), have been shown to tolerate some sequence variability (Michael et al., 1992), and their potential for use in combinatorial strategies has in fact already been exploited in the search for new DNA-binding specificities (Rebar and Pabo, 1994; Jamieson et al., 1994). By shifting our focus away from the biological function, we realized it would be possible to use the zinc-finger scaffold to obtain an all-purpose, conformationally homogeneous α-helical peptide library (Bianchi et al., 1995). Our choice was further corroborated by both NMR studies of several zinc-finger peptides with different sequences which revealed a common, well defined structure (Frankel et al., 1987; Parraga et al., 1988; Lee et al., 1989; Omichinsky et al., 1990, Krizek et al., 1991), and by experiments which showed that folding is preserved in an analog with multiple alanine substitutions (Michael et al., 1992).

The library is depicted in Figure 7.3. The parent sequence chosen is a previously described consensus Cys_2His_2 peptide (CP1) with the following sequence (the residues randomized in the library are in boldface): ProTyrLysCysProGlu-CysGlyLysSerPheSerGlnLys**SerAsp**Leu**ValLys**HisGln**Arg**ThrHisThrGly (Krizek et al., 1991; Kim and Berg, 1993).

The structure is organized around a zinc ion coordinated by two cysteines in the β-sheet and two histidines in the α helix, and further stabilized by the hydrophobic packing of three conserved residues in positions 2, 11 and 17. We decided to randomize five posi-tions, defined as X^1-X^5, with X^5 being the closest to the C-terminus (Arg22 in CP1). These residues were selected because they are the most exposed in the helix (accessibility 25–60%) and all cluster on one side of the structure, making the resulting pharmacophore fully available for interaction with a receptor; the distance between the Cα of X^1 and X^5 is about 12 Å.

Two additional features of zinc fingers were of utmost importance for our design: first, in these peptides metal co-ordination and folding are coupled; for each member of the library the same sequence would be available with the expected pharmacophoric structure (metal-bound form) and with any alternative topography (unfolded, metal-free form), with the result that metal-dependent binding would function as a built-in control against the selection of structurally undefined sequences. Second, since both the CD (Kim and Berg, 1993) and the adsorption spectra of peptide-Co(II) complexes (Krizek et al., 1991) are highly

Figure 7.3: Schematic representation of the zinc-finger peptide library. The peptide backbone is shown in ribbon representation; also shown are the zinc ion (central sphere) and the side-chains of the coordinating cysteine and histidine residues. The selectable positions X^1-X^2-X^3-X^4-X^5 are shown as spheres centered on the side-chain β-carbon; (left) view of the structure approximately perpendicular to the helical axis, (right) view approximately along the helical axis.

characteristic of this fold, we would have an easy means to characterise even a large panel of selected variants.

The α-helical library was produced both as fusion with the pVIII protein of filamentous phage and as soluble peptide mixtures by chemical synthesis (Houghten et al., 1991; Pinilla et al., 1992; Wallace et al., 1994). This is one of the advantages of having chosen a very small domain which is amenable to chemical synthesis: on one hand, it enables us to expand the natural repertoire by inserting non-coded amino acids in the pharmacophore, and on the other, having both libraries available allows simultaneous selection with the same receptor. The resulting cross-validation of the consensus sequences separately derived from each screening effectively reduces the incidence of possible artefacts associated with each method. The two α-helical libraries were screened with a monoclonal IgA reactive against the LPS of the human pathogen Shigella flexneri (IgAC5), and the resulting consensus sequences were indeed very similar; the consensus for the phage library was: X^1 = His, X^2 = Phe, X^3 = Val/Ser, X^4 = Gln, X^5 = His/Arg/Gly, and that for the chemical library was: X^1 = His, X^2 = Phe/Fcl/Leu, X^3 = Ala/Val/His,X^4 = Gln/Ser/Ala, X^5 = His/Arg/Asn/Lys. Cross-validation would define the common consensus as: X^1 = His, X^2 = Phe, X^3 = Val, X^4 = Gln, X^5 = His/Arg (peptides $H^1F^2V^3Q^4H^5$ and $H^1F^2V^3Q^4R^5$). When we synthesized peptides corresponding to the sequences defined by both screenings, all the peptides showed strong, zinc-dependent binding to the IgA. The two cross-validated peptides resulted as being among the most active, with $H^1F^2V^3Q^4H^5$ being the best binder of the whole series.

When analyzed by CD and Co(II)-complex absorption, the structure of all the selected peptides was invariant with respect to the parent zinc finger, reinforcing the notion that selection from this library yields ligands of predetermined structure. We also used high-resolution NMR to study the geometrical arrangement of the side chains of the best binder $H^1F^2V^3Q^4H^5$, and again found that the expected topography was conserved.

Since like for the minibody/IL-6 selection there is no reason to believe that the IgA represents a unique case, the zinc finger library is another general source of ligands with predetermined topography.

7.6 Future Developments: Progressing Toward Non-Peptide Pharmaceuticals

We believe that the ability to produce vast repertoires of affinity-selectable, relatively small polypeptide ligands with a constrained, structurally defined conformation would be a major step nearer to the rational design of non-peptide pharmaceuticals (Dougall et al., 1994 and ref. therein). In this respect, the many advantages offered by phage display technology have been counteracted by the limited tools available to force local conformational features into the selectable peptides. The notion of using a protein as a structure-inducing template has been introduced as one way of tackling this difficulty. We thus decided to target two peptide topologies: the reverse turn and the α-helix, and were able to show that by combining protein engineering and phage display we could successfully build vast repertoires of affinity-selectable homogeneously constrained peptide motifs with novel binding activities. We can foresee two possible courses of action following selection: the information derived from screening can be used to attempt the direct design of the corresponding first generation peptidomimetics, to be then refined by the ever expanding peptide chemist's tool kit; or, the selected molecules showing good affinity for the receptor of interest can be probed for their structures and dynamics by NMR or other biochemical techniques to provide further clues to the conformational requirements for binding; we purposely chose the protein templates small enough to apply this strategy.

Although this approach is conceptually appealing, some issues remain to be solved. For instance, hindrance to correct folding imposed by the selected sequence may become a potential limitation (Lee et al., 1994). Restricting the conformational space available to the peptide sequence inevitably lowers the possibilities of finding suitable ligands: to what extent can be established only by putting the libraries to experimental verification. This is true in particular for the

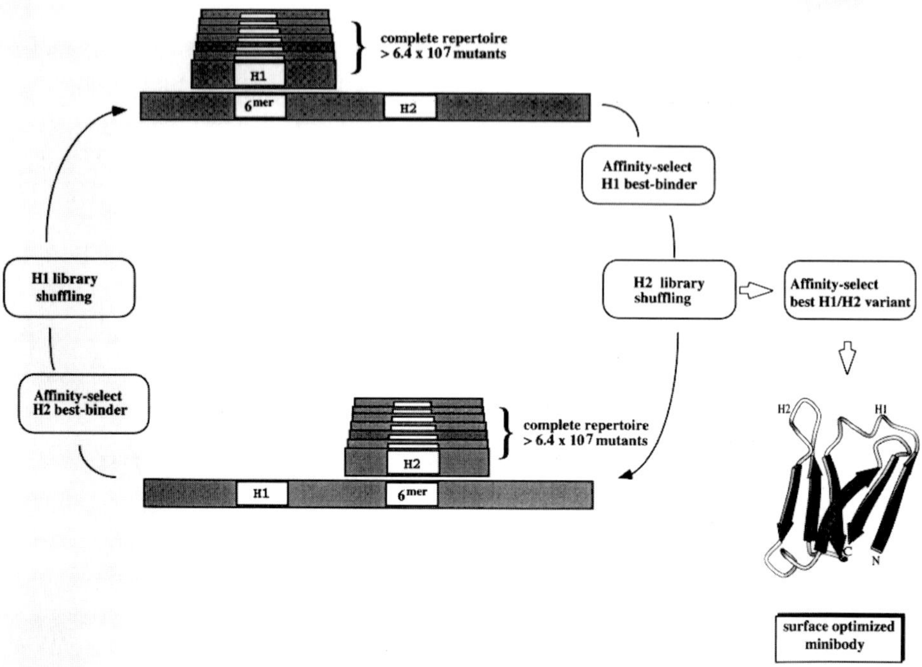

Figure 7.4: Proposed strategy for the surface optimization of the minibody by loop shuffling (see text for explanations).

zinc finger, since contrary to antibodies, a naturally occurring library of α-helical sequences does not exist. It is nevertheless encouraging that so far we have found a carbohydrate mimic utilizing this approach.

The minibody offers an interesting variation on the general strategy so far discussed, because this molecule displays more than one selectable reverse-turn motif. This could enhance the likelihood of finding small polypeptide mimics of complex binding surfaces such as those found in hormone receptors (De Vos et al., 1992) and antigen/antibody interface (Padlan, 1994). Added to the possibility offered by single loop optimization, the presence of two loops enables us to build-up affinity by loop shuffling (Figure 7.4) and subsequent rounds of affinity-selection, as was shown for antibodies (Winter et al., 1994), and work is in progress along these lines.

In conclusion, producing a large repertoire of affinity selectable β-turns and α-helices is a promising step along the path toward the rational design of the low molecular weight non-peptide pharmaceuticals of tomorrow.

Acknowledgements

We are indebted to our colleagues Drs. A. Wallace, M. Nicotra, S. Venturini, F. Martin, A. Folgori, C. Toniatti, G. Ciliberto, R. Bazzo, G. Barbato and A. Tramontano for contributing to various aspects of this work, and J. Clench for reviewing the manuscript.

References

Akira, S., Taga, T. and Kishimoto, T. (1993) Interleukin-6 in biology and medicine, Adv. Immunol. *54,* 1–78.

Barbas, C.F. III, Languino, L.R. and Smith, J.W. (1993) High affinity self-reactive human antibodies by design and selection: targeting the integrin ligand binding site, Proc. Natl. Acad. Sci. USA *90,* 10003–10007.

Beeley, N. (1994) Peptidomimetics and small-molecule drug-design: towards improved bioavailability and *in vivo* stability, TibTech. *12,* 213–216.

Bianchi, E., Venturini, S., Pessi, A., Tramontano, A. and Sollazzo, M. (1994) High level expression and rational mutagenesis of a designed protein, the minibody: from an insoluble to a soluble molecule, J. Mol. Biol. *236,* 649–659.

Bianchi, E., Folgori, A., Wallace, A., Nicotra, A., Acali, S., Phalipon, A., Barbato, G., Bazzo, R., Cortese, R., Felici, F. and Pessi, A. (1995) A Conformationally homogeneous combinatorial peptide library, J. Mol. Biol. *247,* 154–160

Chothia, C. and Lesk, A.M. (1987) Canonical structures for the hyper variable regions of immunoglobulins, J. Mol. Biol. *196,* 901–918.

Chothia, C., Lesk, A.M., Tramontano, A., Levitt, M., Smith-Gill, S.J., Air, G., Sheriff, S., Padlan, E.A., Davies, D., Tulip, W. R., Colman, P.M., Spinelli, S., Alzari, P.M. and Poljak, J. (1989) Conformations of immunoglobulin hypervariable regions. Nature *342,* 877–883.

Davies, J. and Riechmann, L. (1994) 'Camelising' human antibody fragments: NMR studies on VH domains, FEBS lett. *339,* 285–290.

De Vos, A.M., Ultsch, M. and Kossiakoff, A.A. (1992) Human growth hormone and the extracellular domain of its receptor: crystal structure of the complex, Science *255,* 306–312.

Dougall, W.C., Peterson, N.C. and Greene, M.I. (1994) Antibody structure-based design of pharmacological agents, Trends Biotech. *12,* 372–379.

Frankel, A.D., Berg, J.M. and Pabo, C.O. (1987) Metal-dependent folding of a single zinc finger from transcription factor IIIA, Proc. Natl. Acad. Sci. U.S.A. *84,* 4841–4845.Freimuth, P.I., Taylor, J.W. and Kaiser, E.T. (1990) Introduction of guest peptides into Escherichia coli alkaline phosphatase, J. Biol. Chem. *265,* 896–901.

Gallop, M.A., Barrett, R.W., Dower, W.J., Fodor, S.P.A. and Gordon, E.M. (1994) Applications of combinatorial technologies to drug discovery. 1. Background and peptide combinatorial libraries, J. Med. Chem. *37,* 1233–1251.

Hamers-Casterman, C., Atarhouch, T., Muyldermans, S., Robinson, G., Hamers, C., Bajyana Songa, E., Bendahaman, N. and Hamers, R. (1993) Naturally occurring antibodies devoid of light chains, Nature *363,* 446–448.

Hirschmann, R., Nicolau, K.C., Pietranico, S., Savino, J., Leahy, E.M., Sprengler, P.A., Furst, G., Smith, A.B. III, Strader, C. D., Cascieri, M.A., Candelore, M.R., Donaldson, C., Vale, W. and Mechler, L. (1992a) Nonpeptidal peptidomimetics with a β-D glucose scaffolding. A partial somatostatin agonist bearing a close structural relationship to a potent, selective substance P antagonist, J. Am. Chem. Soc. *114,* 9217–9218.

Hirschmann, R., Sprengler, P.A., Kawasaki, T., Leahy, J.W., Shakespeare, W.C. and Smith, A.B. III (1992b) The first design and synthesis of a steroidal peptidomimetic. The potential value of peptidomimetics in elucidating the bioactive conformation of peptide ligands, J. Am. Chem. Soc. *114*, 9699–9701.
Houghten, R.A., Pinilla, C., Blondelle, S.E., Appel, J.R., Dooley, C.T. and Cuervo, J.H. (1991) Generation and use of synthetic peptide combinatorial libraries for basic research and drug discovery, Nature *354*, 84–86.
Hruby, V.J., Al-Obeidi, F. and Kazmiersky, W. (1990) Emerging approaches in the molecular design of receptor-selective peptide ligands: conformational topographical and dynamic consideration, Biochem. J. *268*, 249–262.
Jacobs, G.H. (1992) Determination of the base recognition positions of zinc fingers from sequence analysis, EMBO J. *11*, 4507–4517.
Jameson, B.A., McDonnell, J.M., Marini, J.C. and Korngold, R. (1994) A rationally designed CD4 analogue inhibits experimental allergic encephalomyelitis, Nature *368*, 744–746.
Jamieson, A.C., Kim, S.H. and Wells, J.A. (1994) In vitro selection of zinc fingers with altered DNA-binding specificity, Biochemistry *33*, 5689–5695.
Kahn, M., Wilke, S., Chen, B., Fujita, K., Lee, Y-H. and Johnson, M.E. (1988) The design and synthesis of mimetics of peptide β-turns, J. Mol. Rec. *1*, 75–79.
Kim, C.A. and Berg, J.M. (1993) Thermodynamic β-sheet propensities measured using a zinc-finger host peptide, Nature *362*, 267–270.
Kishimoto, T., Taga, T. and Akira, S. (1994) Cytokine signal transduction, Cell *76*, 253–262.
Kraulis, P.J. (1991) Molscript, a program to produce both detailed and schematic plots of protein structures, J. Appl. Crystallogr. *24*, 946–950.
Krizek, B.A., Amann, B.T., Kilfoil, V.J., Merkle, D.L. and Berg, J.M. (1991) A Consensus zinc finger peptide: design, high affinity metal binding, a pH-dependent structure, and a his to cys sequence variant, Proc. Natl. Acad. Sci. USA. *113*, 4518–4523.
Ku, T.W., Ali, F.E., Barton, L.S., Bean, J.W., Bondinell, W.E., Burgess, J.L., Callahan, J.F., Calvo, R.R., Chen, L., Eggleston, D.S., Gleason, J.G., Huffma, W.F., Hwang, S.M., Jakas, D.R., Karash, C.B., Keenan, R.M., Kopple, K.D., Miller, W.H., Newlander, K.A., Nichols, A., Parker, M.F., Peishoff, C.E., Samanen, J.M., Uzinskas, C.E. and Venslavsky, J.W. (1993) Direct design of a potent non-peptide fibrinogen receptor antagonist based on the structure and conformation of a highly constrained cyclic RGD peptide, J. Am. Chem. Soc. *115*, 8861–8862.
Langen, H.T. and Taylor, J.W. (1992) Alkaline phosphatase somatostatin hybrid proteins as probes for somatostatin -14 receptors, Proteins Struct. Funct. Genet. *14*, 1–9.
Lee, M.S., Gippert, G.P., Soman, K.V., Case, D.A. and Wright, P.E. (1989) Three-dimensional solution structure of a single zinc finger DNA-binding domain, Science *245*, 635–637.
Lee, G., Chan, W., Hurle, M.R., DesJarlais, R.L., Watson, F., Sathe, G.M. and Wetzel, R. (1993) Strong inhibition of fibrinogen binding to platelet receptor $aII_b\beta_{III}$ by RGD sequences installed into a presentation scaffold, Protein Eng. *6*, 745–754.
Luzzago, A., Felici, F., Tramontano, A., Pessi, A. and Cortese, R. (1993) Mimicking of discontinuous epitopes by phage-displayed peptides, I. Epitope mapping of human H ferritin using a phage library of constrained peptides, Gene *128*, 51–57.
Maeda, T., Oyama, R., Ichiara, K., Kimizuka, F., Kato, I., Titani, K. and Sekiguchi, K. (1989) A novel cell adhesive protein engineered by insertion of the Arg-Gly-Asp-Ser tetrapeptide, J. Biol. Chem. *264*, 15165–15168.
Marschall, G.R. (1992) Three-dimensional structure of peptide protein complexes: implications for recognition, Curr. Opinion Struct. Biol. *2*, 904–919.
Martin, F., Toniatti, C., Salvati, A.L., Venturini, S., Ciliberto, G., Cortese, R. and Sollazzo, M. (1994) The affinity selection of a minibody polypeptide inhibitor of human interleukin-6, EMBO J. *13*, 5303–5309.
McDowell, R.S., Blackburn, B.K., Gadeck, T.R., McGee, L.R., Rawson, T., Reynolds, M.E., Robarge, K.D., Somers, T.C., Thorsett, E.D., Tischler, M., Webb, R.R. and Venuti, M.C.

(1994) From peptide to non-peptide. 1. The elucidation of a bioactive conformation of the arginine-glycine-aspartic acid recognition sequence, J. Am. Chem. Soc. *116*, 5077–5083.

McLafferty, M.A., Kent, R.B., Ladner, R.C. and Markland, W. (1993) M13 bacteriophage displaying disulfide-constrained microproteins, Gene *128*, 29–36.

Michael, S.F., Kilfoil, V.J., Schmidt, M.H., Amann, B.T. and Berg, J.M. (1992) Metal binding and folding of a minimalist cys_2his_2 zinc finger peptide, Proc. Natl. Acad. Sci. USA *89*, 4796–4800.

Moore, G.J. (1994) Designing peptide mimetics, Trends Pharm. Sci. *15*, 124–129.

Muyldermans, S., Atarhouch, T., Saldanha, J., Barbosa, J.A.R.G. and Hamers, R. (1994) Sequence and structure of VH domain from naturally occurring camel heavy chain immunoglobulins lacking light chains, Protein Engng . *7*, 1129–1135.

Nolan, W.P., Ratcliffe, G.S. and Rees, D.S. (1992) The synthesis of 1,6-disubstituted indanes which mimic the orientation of amino acid side-chains in a protein alpha-helix motif, Tetrahedron Lett. *33*, 6879–6882.

Olson, G.L., Bolin, D.R., Bonner, M.P., Bös, M., Cook, C.M., Fry, D.C., Graves, B.J., Hatada, M., Hill, D.E., Kahn, M., Madison, V. S., Rusiecki, V.K., Sarabu, R., Sepinwall, J., Vincent, G.P. and Voss, M.E. (1993) Concepts and progress in the development of peptide mimetics, J. Med. Chem. *36*, 3039–3049.

Omichinski, J.G., Clore, G.M., Apella, E., Sakaguchi, K. and Groenborg, A.M. (1990) High-resolution solution structure of the double cys_2his_2 zinc finger from the human enhancer binding protein MBP-1, Biochemistry *29*, 9324–9334.

O'Neil, K.T., Hoess, R.H., Jackson, S.A., Ramachandran, N.S., Mousa, S.A. and DeGrado, W.F. (1992) Identification of novel peptide antagonists for gpIIb/IIIa from a conformationally constrained phage peptide library, Proteins Struct. Funct. Genet. *14*, 509–515.

Padlan, E.A. (1994) Anatomy of the antibody molecule, Molec Immunol. *31*, 169–217.

Parraga, G., Horvath, S.J., Eisen, A., Taylor, W.E., Hood, L., Young, E.T. and Klevit, R. E. (1988) Zinc-dependent structure of a single-Finger domain of yeast ADR1, Science *241*, 1489–1492.

Pavletich, N.P. and Pabo, C.O. (1991) Zinc finger-DNA recognition: crystal structure of a Zif268-DNA complex at 2.1 Å, Science *252*, 809–817.

Pellegrino, G.R. and Berg, J.M. (1991) Identification and characterization of "zinc-finger" domains by the polymerase chain reaction, Proc. Natl. Acad. Sci. USA *88*, 671–675.

Pessi, A., Bianchi, E., Crameri, A., Venturini, S.,Tramontano, A. and Sollazzo, M. (1993) A designed metal-binding protein with a novel fold, Nature *362*, 367–369.

Pinilla, C., Appel, J.R., Blanc, P. and Houghten, R.A. (1992) Rapid identification of high affinity peptide ligands using positional scanning synthetic peptide combinatorial libraries, Biotechniques *13*, 901–905.

Radford, S.E., Dobson, C.M., Evans, P.A. (1992) The folding of hen lysozyme involves partially structured intermediates and multiple pathways, Nature *358*, 302–307.

Rebar, E.J. and Pabo, C.O. (1994) Zinc finger phage: affinity selection of fingers with new DNA-binding specificities, Science *263*, 671–673.

Richardson, J.S., Richardson, D.C., Tweedy, N.B., Gernert, K.M., Quinn, T.P., Hecht, M.H., Erickson, B.W., Yan, Y., McClain, R.D., Donlan, M.E. and Surles, M.C. (1992) Looking at proteins: representations, folding packing, and design. Biophysical society national lecture, Biophys J. *63*, 1186–1209.

Rizo, J. and Gierasch, L.M. (1992) Constrained peptides: models of bioactive peptides and protein structures, Ann. Rev. Biochem. *61*, 387–418.

Roberts, B.L., Markland, W., Ley, A.C., Kent, R.B., White, D.W., Guterman, K. and Ladner, R.C. (1992) Directed evolution of a protein: selection of potent neutrophil elastase inhibitors displayed on M13 fusion phage, Proc. Natl. Acad. Sci. USA *89*, 2429–2433.

Saragovi, H.U., Fitzpatrick, D., Raktabur, A., Nakanishi, H., Kahn, M. and Greene, M.I. (1991) Design and synthesis of a mimetic from an antibody complementarity determining region Science *253*, 792–795.

Saragovi, H.U., Greene, M.I., Chrusciel, R.A. and Kahn, M. (1992) Loops and secondary structure mimetics: development and applications in basic science and rational drug design, BioTech. *10,* 773–778.

Savino, R., Lahm, A., Salvati, A.L., Ciapponi, L., Sporeno, E., Altamura, S., Paonessa, G., Toniatti, C. and Ciliberto, G. (1994) Generation of interleukin-6 receptor antagonists by molecular modeling guided mutagenesis of residues important for gp 130 activation, EMBO J. *13,* 1357–1367.

Sato, M., Lee, J.Y.H., Nakanishi, H., Johnson, M.E. Chrushiel, R.A. and Kahn, M. (1992) Design, synthesis and conformational analysis of γ-turn peptide mimetics of bradykinin, Biochem .Biophys. Res. Commun. *187,* 999–1006.

Schiller, P.W. and DiMaio, J. (1983) In: Peptides: Structure and function (Hruby, V.J. and Rich, D.H.; eds.) pp. 269–278, Pierce Chem. Co., Rockford IL.

Smith, G.P. (1991) Surface presentation of protein epitopes using bacteriophage expression systems, Curr. Opinion Biotechnol. *2,* 668–673.

Smith, A.B. III, Keenan, T.P., Holcomb, R.C., Sprengler, P.A., Guzman, M.C., Wood, J.L., Carroll, P.J. and Hirschmann, R. (1992) Design, synthesis and crystal structure of a pyrrolidone-based peptidomimetic of a β-strand: potential application to the design of novel inhibitors of proteolitic enzymes, J. Am. Chem. Soc. *114,* 10672–10674.

Smith, A.B. III, Hirschmann, R., Pasternak, A., Akaishi, R., Guzman, M.C., Jones, D.R., Keenan, T.P., Sprengler, P.A., Darke, P., Emini, E., Holloway, M.K. and Schleif, W.A. (1994) Design and synthesis of peptidomimetic inhibitors of HIV-1 protease and renin. Evidence for improved transport, J. Med. Chem. *37,* 215–218.

Sollazzo, M., Billetta, R. and Zanetti, M. (1990) Expression of an exogenous peptide epitope genetically engineered in the variable domain of an immunoglobulin: implications for antibody and peptide folding, Protein Engng. *4,* 215–220.

Sun , M., Li , L., Gao, Q.S. and Paul, S. (1994) Antigen recognition by an antibody light chain, J. Biol. Chem. *269,* 734–738.

Taub, R. and Greene, M.I. (1992) Functional validation of ligand mimicry by anti-receptor antibodies: structural and therapeutics implications Biochemistry *31,* 7431–7435.

Tramontano, A., Chothia, C. and Lesk, A.M. (1990) Framework residue 71 is a major determinant of the position and conformation of the second hypervariable region in the VH domains of immunoglobulins, J. Mol. Biol. *215,* 175–182.

Tramontano, A., Bianchi, E., Venturini, S., Martin, F., Pessi, A. and Sollazzo, M. (1994) The making of the minibody: an engineered β-protein for the display of conformationally-constrained peptides, J . Mol . Rec. *7,* 9–24.

Venturini, S., Martin, F. and Sollazzo, M. (1994) Phage display of the minibody: a β-scaffold for the selection of conformationally-constrained peptides, Protein Peptide Letters *1,* 70–75.

Wallace, A., Altamura, S., Toniatti, C., Vitelli, A., Bianchi, E., Delmastro, P., Ciliberto, G. and Pessi, A. (1994) A Multimeric Synthetic Peptide Combinatorial Library, Peptide Res. *7,* 27–31.

Ward, S.E., Güssow, D., Griffiths, A. D., Jones, P. T. and Winter, G. (1989) Binding activities of a repertoire of single immunoglobulin variable domains secreted from *Escherichia coli.,* Nature *341,* 544–546.

Wetzel, R. (1991) Commentary: learning from the immune system; laboratory methods for creating and refining molecular diversity in polypeptides, Protein Engng. *4,* 371–374.

Winter, G., Griffiths, A.D., Hawkins, R.E. and Hoogenboom, H.R. (1994) Making antibodies by phage display technology, Ann. Rev. Immunol. *12,* 433–455.

Wolfson, A.J. Kanaoka, M. Lau, F.T.K. and Ringe, D. (1991) Insertion of an elastase-binding loop into interleukin-1β, Protein Engng. *4,* 313–317.

Yawata, H., Yasukawa, K., Shunji, N., Murakami, M., Hibi, M., Taga, T. and Kishimoto, T. (1993) Stucture-function analysis of human IL-6 receptor: dissociation of amino acid residues required for IL-6 binding and for IL-6 signal transduction through gp 130, EMBO J. *12,* 1705–1712.

Zanetti, M., Filaci, G., Lee, R.H., Del Guercio, P., Rossi, F., Bacchetta, R., Stevenson, F., Barnaba, V. and Billetta, R. (1993) Expression of conformationally constrained adhesion peptide in an antibody CDR loop and inhibition of natural killer cytotoxic activity by an antibody antigenized with the RGD motif, EMBO J. *12*, 4375–4384.

8 Discovery of Disease-Specific Mimotopes by Screening Phage Libraries with Human Serum Samples

Alfredo Nicosia , Paolo Monaci, Allessandra Luzzago, Giovanni Galfré, Franco Felici, Caterina Prezzi, Carmela Mennuni, Annalisa Meola, Monica Mecchia and Riccardo Cortese

8.1 Introduction

Phage-displayed random peptide libraries (RPL) have been extensively used as a source of ligands for several different receptors. A large variety of phage-borne peptide mimics (phagotopes) has been isolated by affinity purification using monoclonal antibodies (mAbs) recognizing linear as well as discontinuous epitopes, and also non proteinaceous molecules (for a review see Cortese et al., 1994; Felici et al., 1995).

The ability of mimicking the natural antigen by phage displayed peptides can be considered at two different levels: a first one is antigenic mimicry, the property of being recognized by the antigen-binding site of the antibody and of competing with the natural antigen for its binding. A second level is immunogenic mimicry: the ability of eliciting *in vivo* an humoral immune response against the natural antigen. Antigenic mimicry is an expected feature of phagotopes selected with antibodies against a specific antigen. As a matter of fact, all the phagotopes identified to date using mAbs have been shown to bind to the antigen-binding site (Cwirla et al., 1990; Felici et al., 1993; Luzzago et al., 1993; Yayon et al., 1993; Motti et al., 1994; Miceli et al., 1994; Hoess et al., 1994; Orlandi et al., 1994). In contrast, immunogenic mimicry of mAb-selected phagotopes is less obvious and not predictable (Keller et al., 1993; Felici et al., 1993; Motti et al., 1994; Orlandi et al., 1994).

The paratope of an antibody is a negative image of the antigen and a peptide mimic (mimotope) specifically binding to it could be a positive image of the natural antigen. This positive image might also be a suitable substitute immunogen, completely distinct from the original antigen. However, in the process of derivation of mAbs, only few paratopes, among the large repertoire of those binding to an epitope, are selected following immunisation of one particular animal. Therefore, as a mAb paratope needs not to be an accurate negative image of the natural epitope, so the phagotopes selected for their best fitting to that particular paratope may not be accurate positive images of the natural epitope. Since the population of paratopes present in immune sera constitutes a more comprehensive and accurate negative image of an antigen, phagotopes are more likely to represent good immunogenic mimics of such antigen, if they are selected with immune sera rather than with a mAb.

In the present paper we present results obtained by using human sera as selector molecules for the isolation of antigenic and immunogenic mimics and describe their potential applications in the field of diagnosis and prevention of infectious and autoimmune diseases

8.2 Using Polyclonal Antibodies as a Ligate for the Selection of RPL

The first example of RPL screening using polyclonal antibodies was reported by Kay and coworkers (1993). After three rounds of affinity selection, four different phage that bound to goat antibodies specific for mouse IgG were identified. Three of these phagotopes displayed a six amino acid-long consensus sequence which was homologous to a peptide present in the mouse γ-2 and γ-3 heavy chains. The antibodies used in this study were taken from experimentally immunized animals and subsequently affinity purified on the immunizing antigen (mouse Fc). Therefore, though polyclonal, these antibodies were reacting only with the target antigen and they could be used as if they were monospecific ligates very much like a mAb.

The scenario is more complex if whole sera are used as ligates. In fact, owing to their polyspecificity, crude sera can select a large and heterogeneous populations of phagotopes (10^6–10^7), most of which are not related to the antigen of interest. Thus, in this case the identification of specific phagotopes requires the use of the natural antigen(s) either before RPL affinity selection to pre-adsorb specific antibodies (Kay et al., 1993) or after library enrichment to identify true mimotopes by competition assays. In either case the use of phage display technology (PDT) would be restricted to those instances in which the pathological antigens are available. To overcome this limitation, we developed a more general strategy for the identifcation of disease-specific phagotopes which avails itself only of clinically characterized sera from patients and normal individuals. A three step protocol was devised: a) affinity selection of RPL with disease-specific sera; b) screening of the enriched phage pool using other patients' sera; c) coun-

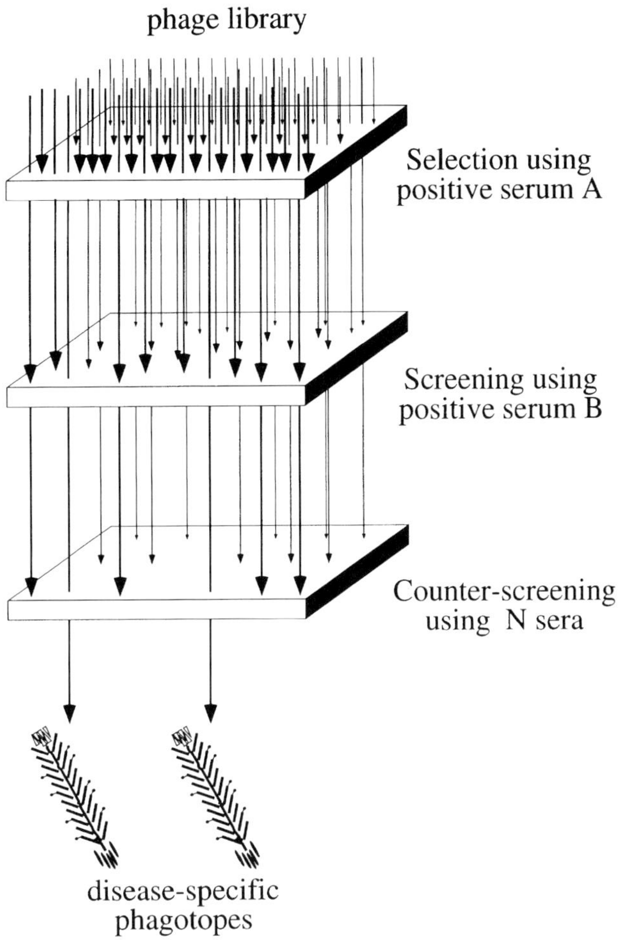

Figure 8.1: Strategy for identifying disease-specific phagotopes using human sera.

ter screening using many different sera from normal individuals (Figure 8.1). The use of pVIII as a carrier for epitope display instead of the more commonly used pIII-based libraries appeared to play a major role in the detection of the reaction between phagotopes and serum antibodies. This is probably due to the much higher multivalence of the former with respect to the latter display system (hundreds vs. three to five peptides per phage particle, respectively; Felici et al., 1991; Luzzago et al., 1993) which increases the avidity of the phage for the binding antibody and results in a higher sensitivity. Following this approach four distinct phagotopes mimicking epitopes of the envelope protein of human hepatitis B virus (HBsAg) were identified using sera from immune and non immune individuals (Folgori et al., 1994). The selected phagotopes were shown to mimic two different epitopes: a first isolate was homologous to amino acids of the *d/y* determinant region of HBsAg which is responsible for viral sub-type specificity (Valenzuela et al., 1980), while the other phagotopes did not show any detectable similarity to the natural antigen.

The possibility of using serum-selected phagotopes as diagnostic reagents was also investigated. The combined use of two non cross-reacting phagotopes detected as positive 80% of the sera from individuals vaccinated with the HBsAg (Folgori et al., 1994). A similar analysis performed using sera from HBV infected patients resulted in 40% of positive score (A. Folgori and F. Felici, unpublished). This last observation allowed to conclude that in the case of HBV infection and probably in other infectious diseases, a larger set of phagotopes should detect the presence of disease-specific antibodies with a high diagnostic sensitivity.

8.3 Immunofingerprint of the Individual Humoral Response to an Infectious Agent: The Hepatitis C Virus

The prediction that a phagotope-based strategy could represent a new and more powerful approach to the diagnosis of viral diseases was confirmed by a study on the human hepatitis C virus (HCV) infection. HCV is the major aetiological agent of non-A non-B hepatitis, a world-wide spread cause of chronic liver disease (Choo et al., 1989). The poor understanding of the humoral immune response to HCV infection, the lack of detailed knowledge of the biology of the virus and the difficulties in expressing viral proteins in a biologically active form render this viral disease an ideal system for applying the PDT in combination with the use of sera to the epitope discovery.

pVIII-based RPLs were affinity selected using sera from clinically well defined HCV-infected patients and subsequently screened for clones reacting with other HCV-positive sera, but does not reacting with sera from healthy controls. In this way several phagotopes were identified (P. Monaci et al., in preparation). In Figure 8.2 the reactivity pattern of some of these phage clones with HCV-positve and negative sera is displayed. Each listed phagotope shows a high frequency of reactivity with HCV-positive sera, but does not react with any of the sera from normal individuals. The difference in the frequency of reactivity between the two populations is statistically very significant, and by this criterium the selected phage clones can be considered disease-specific. However, as described later, this specificity does not necessarily imply that they are all mimics of HCV antigens.

It is important to stress that these phagotopes were selected and identified without using the viral antigens, but requiring only clinically well characterized sets of sera from infected patients and normal individuals.

A number of important features of this system deserves further discussion. The first one is that the selected phagotopes constitute a set of sensitive diagnostic reagents: using just the phage clones listed in Figure 8.2 all tested patients' sera scored positive with at least a subset of them. Should new patients' serum samples score negative with the available set of phagotopes, the same sera could be used as ligates to select new disease-specific phage clones. By iteration, the set of phagotopes can be enlarged to a point approaching 100% sensitivity. Therefore, it is evident that this strategy can have a great impact on the problem of false negative samples.

Extensive search for the maximal number of different HCV-specific phagotopes provided for a large panel of reagents which could very finely dissect the anti-viral humoral response of each individual. This dramatic increase of the number of signals generates a high resolution image of the anti-HCV antibody repertoire present in each patient' s serum. The two principal outcomes of this analysis are the drastic reduction of undeterminate samples and the possibility to "immunofingerprint" the individual humoral response. As shown in Figure 8.2, each serum is characterized by a particular pattern of reactivity. Such typing of the HCV-specific immune response could greatly improve diagnosis and prognosis of the disease.

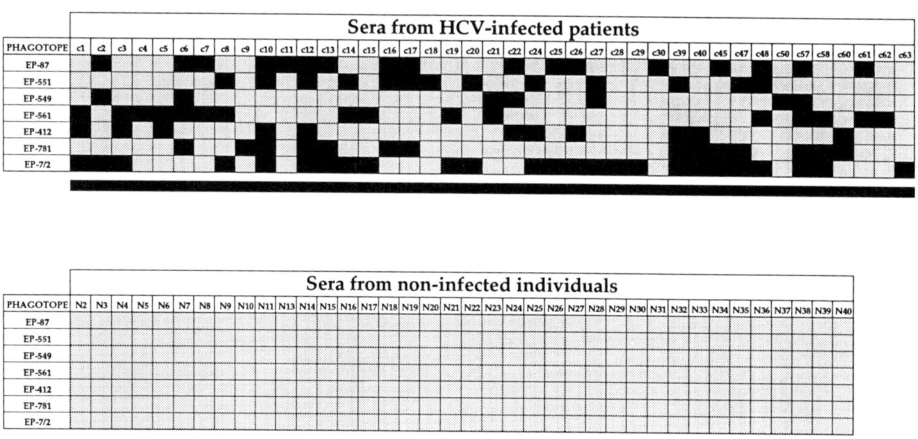

Figure 8.2: Reactivity of HCV-specific phagotopes with human sera. Sera from HCV-infected patients (top panel) and healthy individuals (bottom panel) was measured by ELISA. Positive reactions to the HCV-specific phagotopes (listed on the left of each panel) are represented by black squares; negative reactions are indicated with grey squares.

8.4 Phagotope-Based Vaccines

The use of filamentous phage as immunological carrier was proposed for the first time by de la Cruz and collaborators (1988). They showed that fusion of short peptides derived from the malaria circumsporozoite at the N-terminus of pIII resulted in recombinant phage able to induce specific anti-peptide antibodies. This finding was extended by the group of Perham, who cloned DNA sequences encoding for malaria peptides at the 5′ end of the pVIII gene and obtained good B and T-cell responses (Greenwood et al., 1991; Willis et al., 1993) and by Minenkova et al. (1993) who fused an antigenic determinant of HIV gag protein to pVIII, and used the recombinant phage to successfully immunize rabbits.

Phagotopes selected for their ability to bind mAbs in a specific manner always proved to be good antigenic mimics, but this property was not in every case overlapping with the ability to induce antibodies recognising the natural antigen. In general, when the mAbs utilized for the selections were directed at continuous epitopes, a specific immune-response against the natural antigen was obtained (Keller et al., 1993; Motti et al., 1994). The situation appeared to be more complex when a more elaborate mimicking was requested. Phagotopes selected using a mAb reacting with a discontinuous epitope of the *Bordetella pertussis* toxin failed to induce an anti-toxin response in mice (Felici et al., 1993). Furthermore, two distinct phage clones selected with a mAb against a complex epitope of the HER2/neu oncoprotein were both good antigenic mimics, but only one of them displayed an immunogenic mimicry potential (Orlandi et al., 1994).

Mimotopes displayed as pIII or pVIII fusions and even synthetic peptides reproducing the inserts of mAb-selected phagotopes were able to induce antibodies reacting with the original antigen (Keller et al., 1993; Motti et al., 1994; Orlandi et al., 1994), though no conclusion could be drawn on the relative efficiency of these different forms of presentation of the mimotope to the immune system.

It was of obvious interest to test if disease-specific mimotopes selected using patients' sera, in analogy to anti-idiotype, would be able to induce in naive animals, antibodies ofthe same specificity as those induced by the original antigen, thus closing the pathway of Antigen-Antibody I-Mimotope-Antibody II-Antigen.

The nature of the mimotope themselves, i.e. the kind of epitope (linear or discontinuous) it mimics, and the way it is presented to the immune system, would be expected to play an important role in immunization. In an attempt to define some "immunogenicity rules" which could guide the choice of the best reagents for developing a mimotope-based vaccine, we chose the HBsAg immunization as a model system. This choice relied on the knowledge that most of all antibody response to HBsAg epitopes is known to correlate with protection from viral infection (Waters et al., 1992) and an effective vaccine based on a recombinant protein is already in general use.

As already described we identified HBsAg-specific phagotopes using sera from immune individuals (Folgori et al., 1994). These mimotopes displayed by the selected phage clones were injected in rabbits and in mice in different molecular contexts. Besides using the original phage displaying the mimotope at the N-terminus of pVIII, we prepared other immunogens presenting the mimotopes at the N-terminus of the pIII capsid protein, as fusions to the H chain of human ferritin or to the HBV core peptide and as synthetic multiple antigenic peptides (MAP). The choice of these carriers was mainly dictated by the desire to maintain multimericity of presentation.

In a large set of immunisation experiments we demonstrated that in both rabbits and mice all mimotopes induced HBsAg-specific antibodies (Folgori et al., 1994; Meola et al., 1995). However, in accordance with previous results obtained with genetically engineered phagotopes (Di Marzo Veronese et al., 1994), we found that the strongest and more reproducible responses were obtained with the original phage-displayed mimotopes. This was particularly evident for the mimotopes of discontinuous epitopes.

Direct comparison between the immunogenic mimicry potential of the same mimotope displayed on phage as pIII and pVIII fusion allowed to conclude that the Latter mode of mimotope presentation is more efficient (Meola et al., 1995). The other forms of presentation were in general inferior either because induced lower titers of antibodies or because were not immunogenic in some mouse strains or in the rabbit.

In contrast to the finding of de la Cruz et al. (1988) we found no evidence of genetic restriction of the response in mice. Furthermore, we found that the response to HBsAg following immunisation with phagotopes was as good in a mouse strain known to be non-responder to HBsAg (Milich et al., 1982) as in "high" responder mice (Meola et al., 1995). This finding is best explained by the hypothesis that phage proteins present several T cell epitopes recognized by different mouse strains, and, therefore, the phage acts as efficient carrier even in strains reported to be unable to recognise T epitopes in the HBsAg molecule itself (Milich et al., 1985). A second factor possibly playing an important role, is multimericity of presentation. Anti-HBsAg antibodies can be induced in "low" responders strains by injecting massive doses of the protein (Milich et al., 1982).-ple presentation of the mimotope displayed on pVIII may be a key feature to overcome nonresponsiveness.It has been suggested that the variability of clinical manifestations associated with HBV infection in man could be explained by the range of immune response to HBV antigens (Dudley et al., 1972). Besides, a significant fraction of the human population responds very poorly to HBsAg vaccination. Low responsiveness seems to have a genetic base and association with specific HLA haplotypes has been reported (Alper et al., 1989; Egea et al., 1991; Kruskall et al., 1992). Our finding on phagotope genetically unrestricted response would suggest that effective immunisation with phage could be achieved without the need to identify a priori specific T cell epitopes recognised by the entire population.Another important property of serum-selected phagotopes became evident in studying the fine specificity of mice anti-phagotope sera. One of the HBsAg phage clones was specific for the *d* viral subtype as it was recognised by mice sera immunised against this variant protein, but not by sera from animals injected with the *y* HBsAg. The same phagotope, when used as immunogen, induced antibodies that were not restricted to the *d* variant. In agreement with previous findings with HIV mimotopes (Keller et al., 1993), it may be possible to identify phagotopes that mimic structural features of the original antigen, but are degenerate enough to induce non-subtype specific antibodies. As a matter of fact the mimotope in question showed a significant similarity to the HBsAg (five contiguous amino acids out of nine), but was also different enough that its sequence could not be predicted. Only the screening of large peptide libraries with human sera seems to offer the possibility to identify this type of mimetic molecules and it could be of great importance in developing vaccines against pathogens with a high variability of protective surface epitopes.

8.5 Toward the Identification of the Pathological Antigens of Autoimmune Diseases

The success obtained by applying the phage-display technology to the characterization of the humoral response to known viruses, prompted its application to autoimmune diseases, whose aetiology is largely unknown but might involve infectious agents.

The identification of autoimmune-specific phagotopes using the above mentioned strategy of crossing sera from patients and normal individuals can be a difficult task. Most autoimmune diseases are characterized by a high heterogeneity of the immunological disorder. Very often it is not known whether a defined pattern of antibodies constantly correlates with the disease. The scenario is further complicated by the finding that some autoantibodies associated to a particular disease can be present in patients with other autoimmune syndromes and even in normal individuals, though with a lower incidence. This last issue raises the question of how to define the criteria for classifying the selected phagotopes as specific for the disease.

Dybwad and coworkers (1993) performed multiple rounds of selection/amplification and depletion using alternatively a pool of eighty-four sera from patients with rheumatoidarthritis (RA) and a mixture of sixty sera from normal individuals. In this way, phage reacting with antibodies contributed by different sera, but displaying common specificities, were most efficiently enriched during the selection step. It is not clear whether the depletion step did result in a significant subtraction of phagotopes reacting with normal antibodies, as we have observed that the fraction of phage binding to the antibodies during affinity selection with sera represents only a small percentage of the total number present in the reaction (our, unpublished observations). Nevertheless, these authors were able to select five phage clones which were recognized by patients' sera in ELISA. These phagotopes did not display common motifs. Their association to RA, evaluated by comparing their frequency of recognition by sera from RA patients and normal controls, was significant, but not absolute (44 % vs. 13 %), indicating that humoral response to the selected phagotopes was not an exclusive feature of RA patients.

We also reached similar conclusions by screening RPLs using sera from insulin dependent diabetes mellitus (IDDM) patients. IDDM is a typical example of autoimmunity characterized by a complex multi-factorial pathogenesis and by a genetic susceptibility (Castano and Eisenbarth, 1990). Autoantibodies appear years before the clinical onset of the disease and at the time of the diagnosis, when the pancreatic β-cells are already destroyed, the autoantibodies concentration could be very low or absent making difficult an efficient selection and screening. Therefore for the selection of phagotopes we choose sera from pre-diabetic patients.

Screening several hundreds of affinity selected clones using sera from pre-diabetic and high-risk patients (and from normal individuals as negative controls) identified two phagotopes bearing different, and probably unrelated sequences (Mennuni et al., in preparation). The frequency of reactivity of these phage clones was significantly different between sera from patients and normal controls.

Perhaps the most appealing application of the PDT to the study of autoimmunity is the identification of the endogenous or exogenous targets of the disease-specific immune response. Attempts to use the selected phagotopes as tools for the identification of the IDDM natural autoantigens were made. Serum immunoglobulins were immunopurified using two distinct and non cross-reacting disease related phagotopes as ligates and used for immunohistochemistry experiments. Probing the two pools of purified immunoglobulins on human pancreatic sections showed distinct patterns of reactivity: in one case the insulae and in the second case the pancreatic ducts were positively stained.

The feasibility to apply the PDT to the discovery of autoantigens with patients' sera is further supported by recent data obtained in our laboratory using IgM from HCV-infected cryoglobulinemic patients. Type II mixed cryoglobulinemia (CryoII) is an autoimmune disorder of unknown aetiology which is strongly associated to HCV infection. The main immunological feature of this disease is the presence of cold-insoluble immunocomplexes containing autoantibodies of the IgM class (P.D. Gorevic, 1988). In an attempt to identify the natural autoantigen(s) of these antibodies we aimed at selecting, from a naive repertoire of nonapeptide sequences displayed on phage, the best fitting ones using thenegative image of the antigen imprinted in the binding site of the IgM. A family of phagotopes bearing a common consensus sequence was identified (M. Mecchia et al., in preparation). This sequence was found to be homologous to an exposed region of the humanlymphocyte activation 3 gene product (LAG-3): a membrane protein selectively expressed in activated T lymphocytes and natural killer cells which is structurally and functionally related to CD4 (Triebel et al., 1990: Baixeras et al., 1992; Huard et al., 1994).

We proved that the CryoII IgM phagotopes were true antigenic mimics of LAG-3. Anti-LAG-3 antibodies were induced in rabbits using a synthetic peptide including the region homologous to the selected phage-displayed sequences. These sera specifically recognized both the native protein expressed in induced lymphocytes and the selected phagotopes. Further evidence that the CryoII IgM phagotopes represented the positive image of LAG-3 came from the demonstration that anti-phagotope antibodies raised in rabbits were able to interact in a selective manner with the above LAG-3 peptide. That LAG-3 was indeed the natural target of CryoII IgM was demonstrated by the specific binding of IgM from patients' sera to the LAG-3 peptide. This interaction was inhibited by the peptide itself and by purified phagotope with comparable efficiency.

8.6 Conclusions

The most recent approaches toward the generation of diagnostic reagents and vaccines require detailed information on the pathological antigens. However, in many infectious, allergic and autoimmune diseases the aetiological agent or its antigens are unknown.

An alternative approach is to point straight to the identification of mimic of the determinants of the pathological antigens that are immunologically relevant to the host. This can now be done through the combined use of patients' sera and PDT which allows to define common and peculiar features in the humoral immune response of different patients to the same antigen. The next step will be to verify the range of application of this approach in the fields of diagnosis and prognosis. The preparation of increasingly complex and structurally constrained libraries (O'Neil et al., 1992; Luzzago et al., 1993; McLafferty et al., 1993; Martin et al., 1994; Bianchi et al., 1995) together with the sophistication of the screening methods give good reasons to be optimistic for the PDT-based strategy as being used as a routine method for epitope discovery.

Recombinant molecules are increasingly used to induce a specific immune response to the natural antigen. The successes reported in the use of phagotopes as even better immunogens than synthetic peptides or other carrier proteins holds the promise of using the PDT as a straightforward, inexpensive and widely applicable route to the development of vaccines.

Acknowledgments

We wish to thank G. Bottazzo, U. Di Mario, F. Dotta, M. Casato and M. Fiorilli for providing reagents and for sharing unpublished results. We also thank J. Clench for proofreading the manuscript and M. Emili for artwork.

References

Alper, C.A., Kruskall, M.S., Marcus Bagley, D., Craven, D.E., Katz, A.J., Brink, S.J., Dienstag, J.L., Awdeh, Z. and Yunis, E. J. (1989) Genetic prediction of nonresponse to hepatitis B vaccine, N. Engl. J. Med. *321,* 708712.

Baixeras, E., Huard, B., Miossec, C., Jitsukawa, S., Martin, M., Hercend, T., Auffray, C., Triebel, F. and Piatier-Tonneau, D. (1992) Characterization of the lynphocyte activation gene3-encoded protein. A new ligand for human leukocyte antigen class II antigens, J. Exp. Med. *176,* 327337.

Bianchi, E., Folgori, A., Wallace, A., Nicotra, M., Acali, S., Phalipon, A., Barbato, G., Bazzo, R., Cortese, R., Felici F. and Pessi A. (1995) A conformationally homogeneous combinatorial peptide library, J. Mol. Biol., in press.

Castano, L. and Eisenbarth, G.S. (1990). Type I dyabetes: a chronic autoimmune disease of human, mouse and rat, Annu. Rev. Immunol. *8,* 647679

Choo, Q.L., Kuo, G., Weiner, A.J., Bradley, L.R. and Houghton, M. (1989). Isolation of a cDNA clone derived from a blod-borne non-A non-B hepatitis genome, Science *244,* 359–362.

Cortese, R., Felici, F., Galfre', G., Luzzago, A., Monaci, P. and Nicosia, A. (1994). Epitope discovery using peptide libraries displayed on phage, Tibtech *12,* 262–267.

Cwirla S.E., Peters E.A., Barrett R.W. and Dower W.J. (1990) Peptides on phage: a vast library of peptides for identifying ligands, Proc. Natl. Acad. Sci. *87,* 6378–6382.

de la Cruz, V.F., Lal, A.A. and McCutchan, T.F. (1988). Immunogenicity and epitope mapping of foreign sequences via genetically engineered filamentous phage, J. Biol. Chem. *263,* 43184322.

Di Marzo Veronese, F., Willis, A.E., Boyer-Thomson, C., Appella, E. and Perham, R.N. (1994) Structural mimicry and enhanced immunogenicity of peptide epitopes displsyed on filamentous bacteriophage. The V3 loopof HIV-1 gp120, J. Mol. Biol., in press.

Dudley, F.J., Scheuer, P.J. and Sherlock, S. (1972) Natural history of hepatitis-associated antigen-positive chronic liver disease, Lancet *2,* 1388–1393.

Dybwad, A., Forre, O., Kjeldsen Kragh, J., Natvig, J.B. and Sioud, M. (1993) Identification of new B cell epitopes in the sera of rheumatoid arthritis patients using a random nanopeptide phage library, Eur. J. Immunol. *23,* 3189–3193.

Egea, E., Iglesias, A., Salazar, M., Morimoto, C., Kruskall, M. S., Awdeh, Z., Schlossman, S.F., Alper, C.A. and Yunis, E.J. (1991) The cellular basis for lack of antibody response to hepatitis B vaccine in humans, J. Exp. Med. *173,* 531–538.Felici, F., Castagnoli, L., Musacchio, A., Jappelli, R. and Cesareni, G. (1991). Selection of antibody ligands from a large library of oligopeptides expressed on a multivalent exposition vector, J. Mol. Biol. *222,* 301–310.

Felici, F., Luzzago, A., Folgori, A., and Cortese, R. (1993). Mimicking of discontinuous epitopes by phage-displayed peptides, II. Selection of clones recognized by a protective monoclonal antibody against the Bordetella pertussis toxin from phage peptide libraries, Gene *128,* 21–27.

Felici, F., Luzzago, A., Monaci, P., Nicosia, A., Sollazzo, M. and Traboni, C. (1995). Peptide and protein display on the surface of filamentous bacteriophage, In: Biotechnology Annual Review (El-Geweley, R.; ed.), vol. 1, Elsevier, Amsterdam.

Folgori, A., Tafi, R., Meola, Felici, F., A., Galfrè, G., Cortese, R., Monaci, P. and Nicosia A. (1994). A general strategy to identify mimotopes of pathological antigens using only random peptide libraries and human sera, EMBO J. *13,* 2236–2243.

Gorevic, P.D. (1988). Cryopathies: Cryoglobulins and Cryofibrinogenemia. In Immunological Diseases. Samter, M., Talmage, D.W., Frank, M.M., Austen, K.F. and H.N. Claman editors. 1687–1713.

Greenwood, J., Willis, A.E. and Perham, R.N. (1991) Multiple display of foreign peptides on a filamentous bacteriophage. Peptides from Plasmodium falciparum circumsporozoite protein as antigens, J. Mol. Biol. *220,* 821–827.

Hoess R.0., Mack A.J., Walton H. and Reilly T.M. (1994) Identification of a structural epitope by using a peptide library displayed on filamentous bacteriophage, J. Immunol. *153,* 724–729.

Huard, B., Gaulard, P., Faure, F., Hercend, T. and Triebel, F. (1994) Cellular expression and tissue distribution of the human LAG-3 encoded protein, anMHC class II ligand, Immunogenetics *39,* 213–217.

Kay, B.K., Adey, N.B., He, Y., Manfredi, J.P., Matargnon, A.H. and Fowlkes, D.M. (1993). An M13 phage library displaying random 38-amino acid peptides as a source of novel sequences with affinity to selected targets, Gene *128,* 59–65.

Keller, P.M., Arnold, B.A., Shaw, A.R., Tolman, R.L., van Middlesworth, F., Bondy, S., Rusiecki, V.K., Koenig, S., Zolla Pazner, S., Conard, P., Emini, E.A., and Conley, A.J. (1993). Identification of HIV vaccine candidate peptides by screening random phage epitope libraries, Virology *193,* 709–716.

Kruskall, M.S., Alper, C.A., Awdeh, Z., Yunis, E.J. and Marcus Bagley, D. (1992) The immune response to hepatitis B vaccine in humans: inheritance patterns in families, J. Exp. Med. *175,* 495–502.

Luzzago, A., Felici, F., Tramontano, A., Pessi, A., and Cortese, R. (1993). Mimicking of discontinuous epitopes by phage-displayed peptides, I. Epitope mapping of human H ferritin using a phage library of constrained peptides, Gene *128,* 51–57.

Martin, F., Toniatti, C., Salvati, A.N., Venturini, S., Ciliberto, G., Cortese, R. and Sollazzo, M. (1994) The affinity selection of a minibody polypeptide inhibitor of human interleukin-6, EMBO J. *13,* 5303–5309.

McLafferty, M.A., Kent, R.B., Ladner, R.C. and Markland, W. (1993) M13 bacteriophage displaying disulfide-constrained microproteins, Gene *128,* 29–36.

Meola A., Delmastro P., Monaci P., Luzzago A., Nicosia A., Felici F., Cortese R. and Galfre' G. (1995). Derivation of vaccines from mimotopes: immunologic properties of human hepatitis B virus surface antigen mimotopes displayed on filamentous phage. The Journal of Immunology, in press.

Miceli R.M., DeGraaf M.E. and Fisher H.D. (1994) Two-stage selection of sequences from random phage display library delineates both core residues and permitted structural range within an epitope, Journal Imm. Meth. *167,* 279–287.

Milich, D.R. and Chisari, F.V. (1982) Genetic regulation of the immune response to hepatitis B surface antigen (HBsAg). I. H-2 restriction of the murine humoral immune response to the a and d determinants of HBsAg, J. Immunol. *129,* 320–325.

Milich, D.R., Louie, R.E. and Chisari, F.V. (1985) Genetic regulation of the immune response to hepatitis B surface antigen (HBsAg). V. T cell proliferative response and cellular interactions, J. Immunol. *134,* 4194–4202.

Minenkova, O.O., Ilyichev, A.A., Kishchenko, G.P. and Petrenko, V.A. (1993) Design of specific immunogens using filamentous phage as the carrier, Gene *128,* 85–88.

Motti, C., Nuzzo, M., Meola, A., Galfrè, G., Felici, F., Cortese, R., Nicosia A., and Monaci, P. (1994). Recognition by human sera and immunogenicity of HBsAg mimotopes selected from an M13 phage display library, Gene *146,* 191–198.

O'Neil, K.T., Hoess, R.H., Jackson, S.A., Ramachandran, N.S., Mousa, S.A. and DeGrado, W.F. (1992) Identification of novel peptide antagonists for GPIIb/IIIa from a conformationally constrained phage peptide library, Proteins Struct. Funct. Genet. *14,* 509–515.

Orlandi, R., Ménard, S., Colnaghi, M.I., Boyer, C.M. and Felici, F. (1994). Antigenic and immunogenic mimicry of the HER2/neu oncoprotein by phage-displayed peptides, Europ. J. Immunol. *24,* 2868–2873.

Triebel, F., Jitsukawa, S., Baixeras, E., Roman-Roman, S., Genevee, C., Viegas-Pequignot, E. and Hercend, T. (1990) LAG-3, a novel lymphocyte activation gene closely related to CD4, J. Exp. Med. *171,* 1393–1405.

Valenzuela, P., Quiroga, M., Zaldivar, J., Gray, P. and Rutter, W.J. (1980) Animal Virus genetics, (Fields, B.N., Yaenisch, R. and Fox, C.F.; eds.), pp. 55–77, Academic, New York.

Waters, J.A., Brown, S.E., Steward, M.W., Howard, C.R. and Thomas, H.C. (1992) Analysis of the antigenic epitopes of hepatitis B surface antigen involved in the induction of a protective antibody response, Virus. Res. *22,* 1–12.

Willis, A.E., Perham, R.N. and Wraith, D. (1993) Immunological properties of foreign peptides in multiple display on a filamentous bacteriophage, Gene *128,* 79–83.

Yayon A., Aviezer D., Safran M., Gross J.L., Heldman Y., Cabilly F., Givol D. and Katchalski-Katzir E. (1993) Isolation of peptides that inhibit binding of basic fibroblast rowth factor to its receptor from a random phage-epitope library, Proc. Natl. Acad. Sci. *90,* 10643–10647.

9 The Utilization of Platelets and Whole Cells for the Selection of Peptides Ligands from Phage Display Libraries*

Michael V. Doyle, Laura V. Doyle, Susan Fong, Robert J. Goodson, Lootsee Panganiban, Robert Drummond, Jill Winter, Steven Rosenberg

9.1 Introduction

Large and diverse peptide libraries are most easily synthesized and screened utilizing the phage display technology (Scott and Smith, 1990; Cwirla et al., 1990; Devlin et al., 1990). This technology is distinct from synthetic peptide affinity selections (Geysen et al., 1986; Houghten et al., 1991) in that the selected peptide displaying phage can be amplified via infection of *E. coli,* isolated and reselected. The identity of the peptide is easily obtained by sequencing the DNA genome that accompanies it. Target proteins can be bound to plastic (microtiter wells or falcon dishes) magnetic beads or displayed on cells. We have utilized this technology to select high affinity peptides for both purified and cellular targets (Goodson et al., 1994; Fong et al., 1994). In this report we will discuss 3 targets and the strategies used to successfully identify peptides that bind to these targets.

9.2 Platelets

The affinity selection of phage displayed libraries on platelets was an early step in the development of a new technology (Fong et al., 1994). The aim was to screen for drugs that bound to platelets and which inhibited aggregation. This

* parts of this manuscript are reprinted from Fong, S. Doyle, L.V., Devlin, J.J. and Doyle, M.V. (1994) Drug Dev. Res. 33, 64–70 by permission of Wiley-Liss, a subsidary of John Wiley & Sons, Inc.

approach allows for drug lead discovery based on the availability of a cellular assay as opposed to the availability of purified and characterized components. The feasibility of the approach was demonstrated by the selection of several classes of aggregation inhibitors.

The aggregation of platelets plays a critical role in normal hemostasis and in many pathological disorders caused by inappropriate thrombogenesis. Thus we set out to identify peptide ligands that might affect this function. Furthermore, platelets have the advantage that the previously described inhibition of platelet aggregation by RGD-containing peptides would serve as an internal control.

Intact, plastic adherent platelets were used as the target for affinity selection of the random-peptide, phage-display library. We used the agonist ADP to produce a uniformly activated population of platelets; we have not determined whether similar results could be obtained with non-adherent or non-activated platelets. The random peptides were displayed on an M13-derived filamentous phage as an amino terminal fusion to the minor coat protein pIII, which is present at about five copies per phage particle (Devlin et al., 1990). After incubation of the phage library with platelets, weakly bound phage were removed by extensive washing. The remaining phage were eluted and then amplified (by infection of *E. coli.*) for reselection. There was no observed increase in the yield of phage during the first three rounds of selection, presumably due to a high level of non specific background binding. By the fourth round of selection (Table 9.1), the yield of library phage eluted from platelets had increased ten-fold, suggesting that the population was becoming enriched for phage displaying peptide sequences that bound to platelets. The increased yield was even more convincing when analyzed at a clonal level. Forty-three of 48 randomly chosen clones from round 4 bound to platelets and could be eluted with yields of 1–10%, while the yields of the other 5 clones as well as of 15 randomly selected phage from the original library were less than 0.05%. Yields of 1–10% are typical for binding of a clonal phage to its respective affinity matrix (Scott and Smith, 1990; Christian et al., 1992; Scott et al., 1992). DNA sequence analysis of these 43 platelet-binding phage revealed 17 different predicted fusion peptides (Tables 9.2 and 9.3). Since all the mature pIII fusion proteins of the phage display vector M13LP67 are predicted to begin with an invariant AE dipeptide to facilitate processing of the pIII signal peptide (Devlin et al., 1990), all the sequences shown in Tables 9.2 and 9.3 (as well as the synthetic peptides used in this study) reflect the presence of these two amino acid residues at the amino terminus.

The power and utility of selecting phage-display peptide libraries on intact platelets is evident from the diversity and biological activity of the peptides isolated. These sequences – from a single set of phage selections – revealed a series of different peptide motifs that inhibit platelet aggregation including some that recapitulate and extend the work of a number of different investigators.

Table 9.1: Enrichment of Platelet-Binding Phage by Repetitive Affinity Selection.

Selection Round	Total PFU		% Yield
	Input	Urea Eluate	
1	2.0×10^{10}	2.0×10^{6}	0.010
2	1.0×10^{11}	7.0×10^{6}	0.007
3	1.0×10^{11}	1.4×10^{7}	0.014
4	1.4×10^{11}	2 .0x 10^{8}	0.140

The results of affinity selection are presented as the totl number of pfu present in the input and in the urea eluate for each round.

Table 9.2: Random Petides Displayed by Platelet-Binding Phage: Predicted Amino Acid Sequences and Effects on ADP-Induced Platelet Aggregation.

Peptide Sequence	Motif	IC_{50}, μM
AE HYTFPKHSVWDIPYI	KHS	80 ± 4 (3)
AE QTPKHSLWTPYTYPA	KHS	240 ± 30 (2)
AE AHYKHSCRQDFCPTL	KHS	88 ± 20 (3)
AE KNIVTYIPRGDMPTM	Kistrin	8 ± 2 (4)
AE YKHPRGDSPTHHQFG	Kistrin	19 ± 1 (3)
AE ARIMRGDMPSSNHFT	Kistrin	6 ± 2 (3)
AE PGLAISCKRDVCITS	KRDS	223 ± 5 (2)
AE TQYPHPVCRGDCNMW	RGD	60 ± 13 (3)
AE PRGDWSTFRTHDTLF	RGD	16 ± 3 (3)
AE TLLDQGGHAVQLGDV	Fibrinogen γ chain	N.E. (2)
AE FSKRTSEAFPVWRLM		48 ± 8 (4)
AE ASVRSYLHKRLEHPL		N.E. (2)
AE FYRPSLLTAWYSNMP		N.E. (2)
AE SASFSCRSDYCITST		200 ± 17 (3)

The results of platelet aggregation assays ar presented as the mean ± standard deviation of the IC_{50} velues. IC_{50} is defined as the concentration of synthetic peptide that produced a 50% decrease (IC_{50}) in the initial slope of the light transmission curve compared to the no-peptide control. Numbers in parentheses are the number of independent measurements N.E. means no effect at 250 μM. See text for discussion of the different motifs.

Table 9.3: Random Petides Displayed by Platelet Binding Phage: Predicted Amino Acid Sequences and Effects on Ristocetin-Induced Platelet Aggregation.

Peptide Sequence	**IC_{50}, mM**
AE LFSTHYLAFKEDYSQ	4 ± 1 (7)
AE EKHWLYVPDWTTAP	N.E. (2)
AE EKHWLYSPGLVHGP	N.E. (2)

The results of platelet aggregation assays are presented as the mean ± standard deviation of the IC_{50} values.

Five peptides contained RGD sequences suggesting that these fusion peptides bound to the glycoprotein IIb-IIIa (GPIIb-IIIa). GPIIb-IIIa, which comprises the integrin subunits aIIb and b3, is a receptor for fibrinogen, vitronectin, von Willebrand factor, and fibronectin (Phillips et al., 1988; Phillips et al., 1991). These four ligands all contain an RGD sequence and certain small peptides containing RGD sequences will compete with the binding of these ligands to GPIIb-IIIa (Plow et al., 1985). We therefore performed competition platelet-binding experiments between each of the 17 different phage clones and fibrinogen, one of the RGD containing ligands for GPIIb-IIIa. The yield of each phage in Table 9.2 was reduced more than 90% in the presence of fibrinogen which suggested that these phage do bind GPIIb-IIIa. By contrast, the yield of phage in Table 9.3 was unaffected, suggesting that these phage may bind to receptors other than GPIIb-IIIa.

The presence of the RGD sequence in several of these peptides and their apparent targeting of the GPIIb-IIIa complex indicated that at least some of these peptides might have effects on platelet functions. We therefore tested the ability of all the peptides listed in Tables 9.2 and 9.3 to inhibit platelet aggregation. Because the fibrinogen competition experiments also suggested that some of the peptides did not bind to GPIIb-IIIa, the aggregation inhibition experiments were performed with two different platelet agonists. Activation of platelets with the classical agonist ADP causes aggregation mediated by the GPIIb-IIIa receptor and plasma fibrinogen. The second platelet agonist was the glycoprotein ristocetin which facilitates the interaction of soluble von Willebrand factor with the platelet glycoprotein Ib-IX complex resulting in aggregation or agglutination (Ruggeri, 1991; Ruggeri et al., 1983).

Under conditions of high shear stress, such as occur at the walls of arterial blood vessels, the initial adhesion of platelets to damaged endothelium is believed to be mediated by binding of von Willebrand factor in the extracellular matrix to the glycoprotein Ib-IX complex on platelets, an interaction that does not depend on RGD sequences or GPIIb-IIIa (Ruggeri and Ware, 1993; Sakariassen et al., 1986). A synthetic peptide based on one of the sequences in Table 9.3 had significant inhibitory effects upon the aggregation of platelets induced by ristocetin.

A:

AEHYTFP**KHS**VWDIPYI

AEQTP**KHS**LWTPYTYPA

AEAHY**KHS**CRQDFCPTL

p**KHS**xw

B:

...CRIP**RGD**MPDDRC...

aeknivtylP**RGD**MPtm

aeaRlm**RGD**MPssnhft

aeykhP**RGD**sPthhqfg

Figure 9.1: Inhibitory Peptide Motifs
Inhibitory peptide motifs. *A:* The alignment of these three peptides suggests the inhibitory motif of KHS. The consensus sequence, shown in the last line, also indicates in the lower case characters that the conserved region may extend to the P and W residues. Note that all residues at the "X" position are hydrophobic. *B:* The sequence from residue 45 to residue 57 of the snake venom protein, kistrin, is shown on the top line and the sequences of the three platelet-binding-peptides, which demonstrate potent inhibition of aggregation, are aligned below. The residues which are identical with those in kistrin are capitalized. There is an apparent preference for a basic residue aligned with kistrin at residue 46, and an S or T at residue 54.

By contrast, none of the peptides in Table 9.2 affected ristocetin-induced aggregation (data not shown). Additionally, this peptide from Table 9.3 inhibited ristocetin-induced agglutination of lyophilized, formalin-fixed platelets, which is frequently used as an assay for ristocetin-mediated interaction of von Willebrand factor and glycoprotein Ib-IX in the absence of active platelet metabolism (-George et al., 1984). Similar results were obtained with preparations of this peptide from several sources; and a randomized peptide (EALYLFEKTQSSAYFHD) of the same amino acid composition had no effect on aggregation. Previously described inhibitors of the interaction between glycoprotein Ib-IX and von Willebrand factor include soluble fragments of both the receptor (Vicente et al., 1990) and the ligand (Sugimoto et al., 1991), monoclonal antibodies (Coller et al., 1983; Sixma et al., 1984) and organic molecules such as Evans blue dye (Kirby, 1975) and aurin tricarboxylic acid (Weinstein et al., 1991). The peptide we isolated bears no obvious sequence or structural similarity to any of these other inhibitors. The other two peptides in Table 9.3 had no effect on aggregation and the receptor for these sequences is unknown.

All but three synthetic peptides based on the predicted sequences in Table 9.2 inhibited ADP-induced platelet aggregation with IC_{50}s ranging from 6 to 240μM.

Table 9.4: Fibrinogen and LRGDSP Prevent Phage Bearing Peptide AEHYTFP KHSVWDIPYI from Binding to Platelets*

Concentration (μM)	% inhibition of binding by	
	Fibrinogen	**LRGDSP**
300	N.T.	95
100	N.T.	88
30	93	55
10	87	42
3	93	10
1	80	N.T.
0.3	63	N.T.
0.1	0	N.T.

* % Inhibition was calculated as follows: 100 x {1 – (pfu titer with inhibitor/pfu titer without inhibitor)}. The pfu titer of this phage binding to the tissue culture plastic well (i. e. background binding) was subtracted from all the results prior to the calculation of % inhibition. N. T., not tested; pfu, plaque forming unit.

A positive control hexapeptide, LRGDSP, had an IC_{50} of 190 ± 28μM (n = 6), which agrees with the published IC_{50} of 225 μM for the peptide GRGDS (Dennis et al., 1990). The first three peptide sequences in Table 9.2 define the KHS motif shown in Figure 9.1 A. The results of competitive platelet-binding experiments between a phage from this group, which is predicted to display the peptide AEHXTFPKHSVWDIPYI, and fibrinogen as well as the hexapeptide LRGDSP are shown in Table 9.4. These data suggest that the target for this phage is the GPIIb-IIIa receptor since both fibrinogen and the RGD peptide prevent the phage from binding to platelets. The target for these peptides may be either within the RGD binding site of the GPIIb-IIIa complex or at a site elsewhere on GPIIb-IIIa that is distinct, but conformationally affected by the presence of ligands bound to the RGD site. Spatially distinct, but mutually exclusive binding to GPIIb-IIIa has been suggested to occur between the fibrinogen α chain peptide RGDS and the fibrinogen γ chain peptide LGGAKQAGDV (Bennett et al., 1988). The differences in the dose response curves between fibrinogen and LRGDSP correspond to the approximately 30-fold difference in their binding affinities for GPIIb-IIIa. Additionally, all three KHS-containing peptides inhibited GPIIb IIIa dependent platelet aggregation with IC_{50}s ranging from 80 to 240 μM (Table 9.2).

A second group of three peptides from Table 9.2 exhibit strong sequence similarity to a 7 residue region of the GPIIb-IIIa antagonist, kistrin, a member of a family of potent, RGD containing, platelet aggregation inhibitors from snake venoms (Dennis et al., 1990). The Cys residues at positions 45 and 57 of kistrin

(Figure 9.1 B) are not disulfide bonded to each other, but are linked to Cys residues elsewhere in the polypeptide (Adler et al., 1991). Kistrin's RGD sequence is displayed as an extended region formed between the two bends induced by the prolines at positions 48 and 53 (Adler and Wagner, 1992). Two of the platelet-selected peptides retained this dual proline-bound structure; the exception replaced the first proline with another hydrophobic amino acid, methionine. These peptides demonstrated 10- to 30-fold greater activity than the positive control peptide LRGDSP as inhibitors of platelet aggregation.

The sequence KRDS, derived from human lactotransferrin, has been reported to inhibit platelet aggregation (Mazoyer et al., 1990). The sequence CKRDVC, present in one of the platelet-binding phage peptides, is similar to this tetrapeptide, and represents a third peptide motif in Table 9.2. The presence of flanking Cys residues suggests a possible cyclization around the relevant tetrapeptide. Other workers have shown that cyclized RGD-containing peptides are more potent antagonists than their linear counterparts (Pierschbacher and Ruoslahti, 1987). The CKRDVC peptide had an IC50 of 223 μM, which is similar to the IC_{50} of 350 μM for the lactotransferrin-derived tetramer, KRDS (Mazoyer et al., 1990). The expected RGD motif is represented by the next two peptides listed in Table 9.2, one of which also contains flanking cysteine residues around the RGD sequence. These two peptides had IC_{50} values of 60 and 16 μM. Using purified GPIIb-IIIa to screen a phage-display library of random hexapeptides bounded by Cys residues, O'Neil, et al. (1992) have identified a series of cyclic RGD- and KGD-containing sequences that inhibit platelet aggregation. Koivunen, et al. (1993) have also obtained potentially cyclic, RGD-containing peptides from a phage-display library by selection on the $\alpha_5\beta_1$ integrin.

The fifth peptide motif is evident in the similarity between the peptide AETLLDQGGHAVQLGDV and the dodecapeptide, HHLGGAKQAGDV, from the g chain of fibrinogen. The dodecapeptide inhibits fibrinogen binding to platelets (Kloczewiak et al., 1984), competes with RGD-containing sequences for binding to purified GPIIb-IIIa (Lam et al., 1987), and inhibits platelet aggregation with an IC_{50} of 1 mM (Adelman et al., 1990). Although fibrinogen inhibited phage bearing the dodecapeptide-like peptide from binding to platelets, the synthetic peptide had no inhibitory effect on platelet aggregation when tested at concentrations up to 500 μM. Of the four remaining platelet-binding peptide sequences in Table 9.2, two of the synthetic peptides did inhibit platelet aggregation and could provide additional information on motifs for ligands binding to GPIIb-IIIa.

This approach to the identification of ligands for cellular receptors has a number of significant advantages. Since whole cells are used as the affinity matrix, the receptors are likely to be in their native conformations. Furthermore, although a large variety of receptors are being screened at one time, additional work with purified receptors is only necessary for the receptors that bind interesting new ligands. Thus this method quickly focuses effort on peptides that do

Table 9.5: Activated Platelet Selection of a Cyclic 6 mer Library

Selected Peptide (frequency)			
cswkRGDc	(5)	cmRRDvpc	(1)
cnstRGDc	(4)	cmarkagc	(1)
cnlkRGDc	(2)	cnakKGDc	(1)
cnrsRGDc	(1)	cnslKGDc	(1)
crniRGDc	(1)	cwRRDvqc	(6)
chwrRGVc	(1)	clRGiDwc	(3)
cmaRQGvc	(1)	crnrKGDc	(6)
		cqwlssgc	(1)

The affinity selection of a phage displayed cyclic 6 mer library with activated platelets resulted in a predominance of RGD clones. Interesting variants such as RGDiD were also obtained.

bind to molecules on the target cells. If the ligands demonstrate activity in based assays, they can then be used as affinity matrices to purify previously undefined receptors. Additionally, although the phage-display library in these experiments had a complexity of 3×10^7, libraries of significantly greater diversity (e.g. 10^8 to 10^9) could be used with no significant increase in effort because screening and selection are separated from the labor-intensive functional analysis. One shortcoming of the phage-display approach to drug discovery is the amount of effort required to clone, perform DNA sequencing, synthesize and test the activity of the peptides from large numbers of selected, target-binding phage. We sequenced 43 platelet-binding phage from round 4 of our selections and obtained 17 different sequences. Further analysis of additional selections from this random peptide, phage-display library would certainly yield other active peptide sequences. Table 9.5 shows sequences derived from activated platelet panning of a cyclic 6 mer library.

We are optimistic that many of the peptides recovered from such selections on whole cells will bind to receptors. Since many of our peptides appear to bind to GPIIb-IIIa, our success at finding receptor binding peptides could be attributed to the high density of GPIIb-IIIa on platelets. However, we have isolated an initial series of peptide-bearing phage that bind to lymphocytes, and these phage do not bind to fibroblasts (unpublished data). These results suggest that several rounds of affinity selection with whole cells results in a set of phage that bind to specific receptors rather than to structural molecules present on many cell types. Furthermore, in previous studies the phage isolated by screening phage-display libraries with purified ligand-binding molecules were shown to compete with known ligands for binding, suggesting that the peptide-bearing phage bound to the ligand binding site (Devlin et al., 1990; Scott and Smith, 1990; Cwirla et al., 1990). Thus, empirically, it seems that affinity isolated, peptide-bearing phage are more likely to bind to a ligand-binding site than to a structural portion of a molecule. There is a critical receptor density on cells required for peptide-bearing phage binding and selection. We have used transfected cells expressing high levels of the human urokinase plasminogen activator receptor (UPAR) to isolate peptides with lownM affinity for UPAR from this random-peptide, phage-display library (Goodson et al., 1994). The motifs discovered by these cell-based approaches serve as lead compounds from which to develop more potent and selective inhibitors. Such studies could include further application of phage-display technology. For example, specific amino acid positions within binding peptides could be randomly mutated, expressed on the phage pIII protein and the resulting population reselected to determine the effects of amino acid substitutions at those positions on binding and on activity.

In summary, we have identified a series of peptide motifs that inhibit platelet aggregation by selecting platelet-binding phage from a random-peptide library on intact, activated platelets. This approach – using whole cells as targets for random-peptide libraries displayed on filamentous phage – has the potential for making important contributions to the identification of pharmacologically active peptide ligands even in the absence of the molecular characterization of their receptors.

9.3 Urokinase Plasminogen Activator Receptor

The urokinase plasminogen activator receptor (UPAR) is an important mediator of metastasis and angiogenesis (Crowley et al., 1993; Fazioli and Blasi, 1994), due to its role in localizing a proteolytic cascade on the surface of invasive cells. Screening of phage display peptide libraries for UPAR ligands was hampered initially by the lack of purified soluble protein, as the native molecule is linked to the membrane via a glycosylphosphatidyl inositol moiety. Thus it was decided to screen against cells displaying UPAR. The human UPAR cDNA was cloned and expressed in baculovirus infected SF9 cells and transfected COS7 cells in preparation for cellular panning experiments (Goodson et al., 1994; Roldan et al., 1990). Since residues 13–32 of uPA had been shown to bind UPAR (Apella et al., 1987), a positive control phage encoding this sequence at the amino terminus of the pIII protein of an M13 derivative, LP67 (Devlin et al., 1990) was made. This positive control phage showed enrichment specifically on SF9 cells expressing UPAR, as shown in Table 9.6.

Table 9.6: Bacteriophage Display of a Known UPAR Binding Petide and a Trial Panning

% Recoveries

Cells Panned	**13–32 Phage**	**M13**
FLUPAR	6.8 %	.009 %
SPR	.008 %	.01 %

The positive control phage display AA 13–32 of UPA at the N Terminus of pIII. The M13 negative control phage yield blue plaques. The positive control is enriched 800 fold on SF9 cells displaying Full Length Urokinase Plasminogen Activator Receptor (FLUPAR) compared to SF9 cells displaying Substance P Receptor (SPR).

Table 9.7

Sequences that Bind SF9 Cells

PQGWWRNEYW	WCGLQLCSNP
SRTWDLNPWW	TISASWSKWY
DGIWRHNEYW	IPWCTPLILTEKPCL
SSPWALNPYW	LLYWPPRNSVGYDPL
CYATWWCEED	LSPPNSMREFITMFY
RSYDCVGFGC	

The sequences were selected by SF9 cells displaying a variety of target receptors. Upon micropanning analysis phage displaying these sequences were shown to bind SF9 cells that lacked the target of interest. The selected 10 mers are from the Christian et al. (1992) library, the 15 mers from the Devlin et al. (1990) library.

Preliminary studies with baculovirus infected SF9 cells indicated that background binding to these cells could pose a problem. Table 9.7 shows sequences that bind to SF9 cells independent of the displayed target . Depletion of these cell binders by a preliminary exposure to wildtype Baculovirus infected cells was attempted but with limited success. It is difficult to deplete a library entirely of the background binding sequences in one round. Instead, a different strategy was developed in which the target was displayed on 2 different cell types and these cell types panned alternately. This approach changes the background while maintaining the target and thus forces a stronger selection for target binding sequences.

As shown in Table 9.8, the alternate panning of UPAR displaying SF9 and COS cells resulted in a dramatic enrichment for binding phage within 3 rounds. Sequence analysis of these phage revealed a predominance in the frequency of

Table 9.8: Affinity Selection of the 15 Mer Library with UPAR Expressing SF9 and COS Cells*

Round	Cells	Input	Output	% Recovery
ONE**	SF9	2.5×10^{10}	3.3×10^{5}	.0013
TWO	COS	3.1×10^{11}	1.2×10^{8}	.039
THREE	SF9	2.8×10^{10}	1.5×10^{9}	5.40

*48 hour post infection or transfection of cells. 1×10^{6} SF9 cells 5×10^{5} COS cells per paning.

Table 9.9: Phage Selected from UPAR Panning of 15mer Library

CLONE	SEQUENCE	IC50 μM	FREQUENCY*
20	PMPHSLNFSQYLWYT	0.01	11
26	HTYSSLWDTYSPLAF	0.34	8
54	LDLWMRHYPLSFSNR	0.38	1
16	SSLWTRYAWPSMPSY	0.40	5
12	WHPGLSFGSYLWSKT	0.40	6
18	PALLNWSFFFNPGLH	1.00	1
9	WSFYNLHLPEPQTIF	1.00	2
11	PLDLWSLYSLPPLAM	2.00	2
42	PTLWQLYQFPLRLSG	2.50	1
48	ISFSELMWLRSTPAF	5.00	1
75	LSEADLWTTWFGMGS	7.00	1
17	SSLWRIFSPSALMMS	8.00	1
13	SLPTLTSILWGKESV	8.00	1
10	TLFMDLWHDKHILLT	8.00	4
44	ILNFPLWHEPLWSTE	9.00	2
14	SQTGTLNTLFWNTLR	10.0	8
38	IKTDEKMGLWDLYSM	23.0	1
19	MHRSLWEWYVPNQSA	>23.0	9
36	SHIKSLLDSSTWFLP	>47.0	1

Phage were selected as described in Goodson et al. (1994). Petide sequences were determined by translation of DNA sequencing results from single-stranded phage templates.

* Number of separate times a given DNA and amino acid sequence was obtained from randomly picked plaques.

+ Apparent inhibition constant of the synthetic peptide or UPA frament for the UPAR-ATF interaction

Table 9.10: Sequence Similarities between Phage Derived SUPAR Binding Petide Sequences

Sequence Name	Sequence	IC50 (μM)
	Set 1	
20	AEPMPHSLN**F**SQ**YLW**YT	0.01
12	AEWHPGLS**F**GS**YLW**SKT	0.40
48	AEIS**F**SE**LMW**LRSTPAF	5.0
	Set 2	
26	AEHTYSS**LW**DT**Y**SPLAF	0.34
54	AELD**LW**MR**H**YPLSFSNR	0.38
16	AESS**LW**TR**Y**A**W**PSMPSY	0.40
18	AEPALLN**W**SF**F**FNPGLH	1.0
9	AEWSF**Y**NL**H**LPEPQTIF	1.0
11	AEPLD**LW**SL**Y**SLPPIAM	2.0
42	AEPT**LW**QL**Y**QFPLRLSG	2.5

The highest-affinity peptide sequences derived from the bacteriophage display library selection are listed. The sequences are divided into two subsets, which have sequence motifs of FXXYLW and LWXXAr (Ar = Y, F,. H, or W). Residues in boldface are conserved within or between subsets.

some clones over others (Table 9.9). Using synthetic peptides we were able to show that they competed with the phage for UPAR binding and competed with each other for the UPAR binding site. Radioreceptor binding assays utilizing biotinylated soluble UPAR and the amino terminal fragment of uPA (ATF) were then performed to determine if the peptides competed with ATF. The IC_{50} values obtained are shown in Table 9.9 and are in the range of 10 nM to 47 uM.

Once biotinylated soluble UPAR was available, affinity selections were also performed utilizing this purified protein rather than the alternating cell selection protocol. Sequences similar to those of Table 9.9 were obtained after 3 rounds of selection on the purified protein.

As shown in the Table 9.10 many of the clones have an LW motif. Analyses are in progress to determine whether there are structural similarities (which are not obvious at the primary sequence level) which account for the competitive binding of the selected peptides and ATF for UPAR.

UPAR is expressed naturally on many cell types, however these cells were not suitable for the panning experiments. In some cases, uPA is also produced by the UPAR displaying cell, which may make the receptor less accessible for panning, or the cells simply did not bind the positive control phage well. In the latter case the amount of UPAR may be lower than that on the SF9 or COS cells, or other unknown characteristics of the cell type may have made panning less efficient. Cells that secrete proteases or have halos of charged or viscous outer matrices, may require extensive optimization for panning. There is even one report in the literature claiming that HEp-2 cells actually ingest phage (Hart et al., 1994).

9.4 Fibroblast Growth Factor Receptor 1

The fibroblast growth factor receptor 1 (FGFR-1) is glycosylated in SF9 cells and thus easily attached to lentil lectin beads. However, an initial panning of libraries with lentil lectin bound FGFR resulted in a predominance of lentil lectin binding phage. The alternate panning strategy defined above was applied to these experiments and phage were alternately panned on FGFR displaying SF9 cells or FGFR-lentil lectin beads. The 26mer library was used in these pannings. There was enrichment over backround, yet no consensus sequences were obtained. In spite of the absence of a motif, several sequences which were isolated at high frequency were tested for their ability to bind FGFR. At least 2 of these sequences bound well as phage and one was later shown to have an IC_{50} of 500nM in a radiolabeled basic FGF competition assay. Neither sequence bears any obvious resemblance to the primary sequence of the natural ligand. The FGFR experiment is yet another application of the alternate panning strategy, emphasizing the utility of subtracting background binders by removing their targets with each round of panning.

Recently, two publications from Genentech have described a strategy for competition panning. In one case, Cunningham, et al. (1994) used phage display to remove cross reactivity from a peptide that binds 2 receptors. Atrial natriuretic peptide (ANP) was displayed on phage and randomized. It was then selected with the receptor of choice, ANP A receptor, while the undesirable receptor was in the "sorting buffer". After several rounds of such competitive panning, ANP A receptor specific variants of ANP were identified. Dennis and Lazarus (1994) used a similar strategy to select Kunitz domain protease inhibitors that would be specific for Tissue factor VIIa while no longer binding the Xa serine protease. Ideally one could utilize such a strategy for the growth factor receptors in order to isolate an agonist or antagonist that is specific for one member of a class of receptors.

References

Adelman, B., Gennings, C., Strony, J. and Hanners, E. (1990) Synergistic inhibition of platelet aggregation by fibrinogen related peptides, Circ. Res. *67*, 941–947.

Adler, M., Lazarus, R.A., Dennis, M.S. and Wagner, G. (1991) Solution structure of kistrin, a potent platelet aggregation inhibitor and GPIIb-IIIa antagonist, Science *253*, 445–448.

Adler, M. and Wagner, G. (1992) Sequential 1H NMR assignments of kistrin, a potent platelet aggregation inhibitor and glycoprotein IIb-IIIa antagonist, Biochemistry *31*, 1031–1039.

Appella, E., Ullrich, S.J., Stoppelli, M.P., Corti, A., Cassani, G. and Blasi, F. (1987) The receptor-binding sequence of urokinase: A biological funtion for growth-factor module of proteases, J. Biol. Chem. *262*, 4437–4440

Bennett, J.S., Shattil, S.J., Pomer, J.W. and Gartner, T.K. (1988) Interaction of fibrinogen with its platelet receptor. Differential effects of a and g chain fibrinogen peptides on the glycoprotein IIb-IIIa complex, J. Biol. Chem. *263*, 12948–12953.

Christian, R.B., Zuckermann, R.N., Kerr, J.N., Wang, L. and Malcolm, B.A. (1992) Simplified methods for construction, assessment and rapid screening of peptide libraries in bacteriophage, J. Mol. Biol. *227*, 711–718.

Cwirla, S.E., Peters, E.A., Barrett, R.W. and Dower, W.J. (1990) Peptides on phage: a vast library of peptides for identifying ligands, Proc. Natl. Acad. Sci. USA *87*, 6378–6382.

Coller, B.S., Peerscke, E.I., Scudder, L.E. and Sullivan, C.A. (1983) Studies with a murine monoclonal antibody that abolishes ristocetin-induced binding of von Willebrand factor to platelets: Additional evidence in support of GPIb as a platelet receptor for von Willebrand factor, Blood *61*, 99–110.

Crowley, C.W., Cohen, R.L., Lucas, B.K., Liu, G., Schuman, M.A. and Levison, A.D. (1993) Prevention of metastasis by inhibition of the urokinase receptor, Proc. Natl. Acad. Sci. USA *90*, 5021–5025

Cunningham, B.C., Lowe, D.G., Li, B., Bennett, B.D. and Wells, J. A. (1994) Production of an atrial peptide variant that is specific for type A receptor, EMBO J. *13*, 2508–2515

Dennis, M.S., Henzel, W.J., Pitti, R.M., Lipari, M.T., Napier, M. A., Deisher, T.A., Bunting, S. and Lazarus, R.A. (1990) Platelet glycoprotein IIb-IIIa protein antagonists from snake venoms: Evidence for a family or platelet-aggregation inhibitors, Proc. Natl. Acad. Sci. USA *87*, 2471–2475.

Dennis, M.S. and Lazarus, R.A. (1994) Kunitz domain inhibitors of tissue factor-factor VIIa, J. Biol. Chem. *269*, 22129–22136.

Devlin, J.J., Panganiban, L.C. and Devlin, P.E. (1990) Random peptide libraries: a source of specific protein binding molecules, Science *249*, 404–406.

Fazioli, F. and Blasi, F. (1994) Urokinase-type plasminogen activator and its receptor: New targets for anti-metastatic therapy? Trends Pharmacol. Sci. *15*, 25–29.

Fong, S., Doyle, L.V., Devlin, J.J. and Doyle, M.V. (1994) Scanning whole cells with phage-display libraries: identification of peptide ligands that modulate cell function, Drug. Dev. Res. *33*, 64–70.

George, J.N., Nurden, A.T. and Phillips, D.R. (1984) Molecular defects in interactions of platelets with the vessel wall, N. Engl. J. Med. *311*, 1084–1098.

Goodson, R.J., Doyle, M.V., Kaufman, S.E., Rosenberg, S. (1994) High-affinity urokinase receptor antagonists identified with bacteriophage peptide display, Proc. Natl.Acad. Sci. USA *91*, 7129–7133

Hart, S.L., Knight, A.M., Harbottle, R.P., Mistry, A., Hunger, H., Cutler, D.F., Williamson, R. and Coutelle, C. (1994) Cell Binding and internalization by filamentous phage displaying a cyclic Arg-Gly-Asp-containing peptide, J. Biol. Chem. 269, 12468–12474.

Houghton, R.A., Pinilla, C., Blondelle, S.E., Appel, J.R., Dooley, C.T. and Cuervo, J.H. (1991) Generation and use of synthetic peptide combinatorial libraries for basic research and drug discovery, Nature *354*, 84–86

Kirby, E.P. (1975) Evans Blue: A specific inhibitor of factor VIII-induced platelet agglutination, Thromb. Haemost. *34*, 770–779.

Kloczewiak, M., Timmons, S., Lukas, T.J. and Hawiger, J. (1984) Platelet receptor recognition site on human fibrinogen. Synthesis and structure-function relationship of peptides corresponding to the carboxy-terminal segment of the g chain, Biochemistry *23*, 1767–1774.

Koivunen, E., Gay, D.A. and Ruoslahti, E. (1993) Selection of peptides binding to the a_5b_1 integrin from phage display library, J. Biol. Chem. *268*, 20205–20210.

Lam, S.C-T., Plow, E.F., Smith, M.A., Andrieux, A., Ryckwaert, J J., Marguerie, G. and Ginsberg, M.H. (1987) Evidence that arginyl-glycyl-aspartate peptides and fibrinogen g chain peptides share a common binding site on platelets, J. Biol. Chem. *262*, 947–950.

Mazoyer, E., Levy-Toledano, S., Rendu, F., Hermant, L., Lu, H., Fiat, A-M., Jolles, P. and Caen, J. (1990) KRDS, a new peptide derived from human lactotransferrin, inhibits platelet aggregation and release reaction, Eur. J. Biochem. *194*, 43–49.

O'Neil, K.T., Hoess, R.H., Jackson, S.A., Ramachandran, N.S., Mousa, S.A. and Degrado, W.F. (1992) Identification of novel peptide antagonists for GP IIb/IIIa from a conformationally constrained phage peptide library, Proteins *14*, 509–515.

Phillips, D.R., Charo, I.F., Parise, L.V. and Fitzgerald, L.A. (1988) The platelet membrane glycoprotein IIb-IIIa complex, Blood *71*, 831–843.

Phillips, D.R., Charo, I.F. and Scarborough, R.M. (1991) GPIIb IIIa: The responsive integrin, Cell *65*, 359–362.

Pierschbacher, M.D. and Ruoslahti, E. (1987) Influence of stereochemistry of the sequence Arg-Gly-Asp-Xaa on binding specificity in cell adhesion, J. Biol. Chem. *262*, 17294–17298.

Plow, E.F., Pierschbacher, M.D., Ruoslahti, E., Marguerie, G.A. and Ginsberg, M.H. (1985) The effect of Arg-Gly-Asp-containing peptides on fibrinogen and von Willebrand factor binding to platelets, Proc. Natl. Acad. Sci. USA *82*, 8057–8061.

Roldan, A.L., Cubellis, M.V., Masucci, M.T., Behrendt, N., Lund, L.R., Dano, K., Appella, E. and Blasi, F. (1990) Cloning and expression of the receptor for human urokinase plaminogen activator, a central molecule in cell surface, plasmin dependent proteolysis, EMBO J. *9*, 467–474

Ruggeri, Z.M., De Marco. L., Gatti, L., Bader, R. and Montgomery, R.R. (1983) Platelets have more than one binding site for von Willebrand factor, J. Clin. Inv. *72*, 1–12.

Ruggeri, Z.M. (1991) The platelet glycoprotein Ib-IX complex, Prog. Hemost. Thromb. *10*, 35–68.

Ruggeri, Z.M. and Ware, J. (1993) von Willebrand Factor, FASEB *7*, 308–316.

Sakariassen, K.S., Nievelstein, P.F., Coller, B.S. and Sixma, J.J. (1986) The role of platelet membrane glycoproteins Ib and IIb-IIIa in platelet adherence to human artery subendothelium, Br. J. Haematol. *63*, 681–691.

Scott, J.K. and Smith, G.P. (1990) Searching for peptide ligands with an epitope library, Science *249*, 386–390.

Scott, J.K., Loganathan, D., Easley, R.B., Gong, X. and Goldstein, I.J. (1992) A family of concanavalin A-binding peptides from a hexapeptide epitope library, Proc. Natl. Acad. Sci. USA *89*, 5398–5402.

Sixma, J.J., Sakariassen, K.S., Stel, H.V., Houdijk, W.P.M., In der Maur, D.W., Hamr, R.J., De Groot, P.G. and van Mourik, J.A. (1984) Functional domains of von Willebrand factor. Recognition of discrete tryptic fragments by monoclonal antibodies that inhibit interaction of von Willebrand factor with platelets and with collagen, J. Clin. Inv. *74*, 736–744.

Sugimoto, M., Ricca, G., Hrinda, M.E., Schreiber, A.B., Searfoss, G.H., Bottini, E., and Ruggeri, Z.M. (1991) Functional modulation of the isolated glycoprotein Ib binding domain of von Willebrand factor expressed in Escherichia coli, Biochemistry *30*, 5202–5209.

Vicente, V., Houghten, R.A. and Ruggeri, Z.M. (1990) Identification of a site in the a chain of platelet glycoprotein Ib that participates in von Willebrand factor binding, J. Biol.

Chem. *265*, 274–280.
Weinstein, M., Vosburgh, E., Phillips, M., Turner, N., Chute Rose, L. and Moake, J. (1991) Isolation from commercial aurintricarboxylic acid of the most effective polymeric inhibitors of von Willebrand factor interaction with platelet glycoprotein Ib. Comparison with other polyanionic and polyaromatic polymers, Blood *78*, 2291–2298.

10 Identification of MHC Binding Motifs with Synthetic and Phage Displayed Peptide Libraries

Juergen Hammer and Francesco Sinigaglia

10.1 Introduction

The antigen-specific immune response is initiated by $CD4^+$ T lymphocytes. They are activated by antigenic fragments (peptides) bound to a relevant major histocompatibility complex (MHC) class II molecules on the surface of antigen-presenting cells. Thus, the prerequisite of any antigen to be recognised by the immune system is its fragmentation by the intracellular processing machinery and the capability of the generated peptides to bind to MHC molecules.

MHC class II molecules consist of a 33-kDa α-chain that is noncovalently associated with a 28 kDa β-chain, both chains are glycosylated transmembrane proteins. Class II molecules are among the most polymorphic proteins known, and each gene has many alleles. Except for the DR α-chain, all the chains are polymorphic. The DR-β gene is the most polymorphic with over 60 alleles. Each allelic form of MHC molecules binds a large array of different peptides to ensure T cell-mediated immunity to the antigenic universe. Until recently, the basis for peptide binding and the structural features common to peptides bound to particular allelic forms of MHC class II molecules remained poorly understood. The use of molecular repertoires such as peptide libraries displayed on the surface of M13 bacteriophage or synthetic designer peptide libraries permitted us to determine the rules which govern MHC class II peptide interactions.

The purpose of this chapter is to review the application of molecular repertoires to elucidate the mechanisms underlying peptide-class II molecule interaction.

10.2 Identification of MHC Class II Peptide Binding Motifs

We have investigated the characteristics of peptides that bind to MHC class II HLA-DR molecules by using large, highly diverse peptide libraries expressed on the surface of bacteriophage M13. This powerful technique is based on the ability of filamentous bacteriophage to display foreign peptides on their outer surface, and involves the screening and enrichment of phage displaying peptides that bind to a particular protein (Devlin et al., 1990; Cwirla et al., 1990; Scott and Smith, 1990).

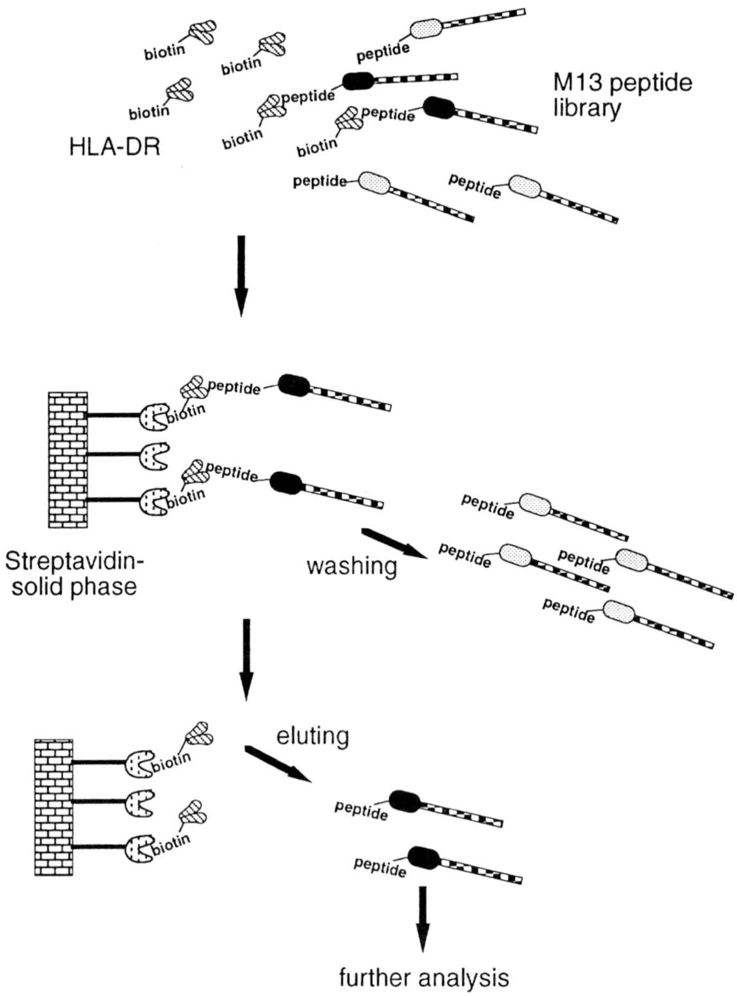

Figure 10.1: Schematic representation of the M13 peptide library screening procedure.

We have inserted oligonucleotides encoding peptides known to bind to the human MHC class II DRB1*0401 molecule into the protein III encoding gene of the bacteriophage M13. Phage expressing the appropriate peptide can bind specifically to the DR molecule (Hammer and Sinigaglia, in press). A library of 20 million random nonamer peptides displayed by M13 was subsequently screened for binding to affinity purified DRB1*0401 molecules. Figure 10.1 summarises the principles of screening the M13 display library. The phage from the library were allowed to bind to biotinylated, immunoaffinity purified and detergent solubilised DR molecules. Bound bacteriophage were in turn attached to a streptavidin-solid phase, and free phage were washed away. The bound phage were then

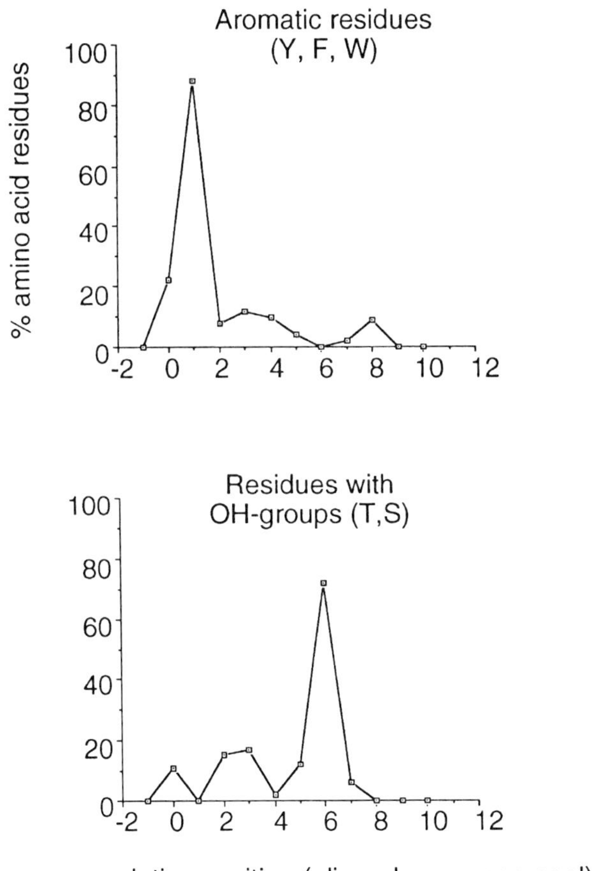

Figure 10.2: Example: identification of major anchor positions i and i+5 within the HLA-DRB1*0401 selected peptide pool.

eluted with acid and amplified by growth in bacterial cells. Subsequent rounds of screening and amplification resulted in a significant enrichment of bacteriophage displaying peptides that bound to the DRB1*0401 molecule.

To investigate the structural characteristics of peptides capable of binding to DR molecules, the peptide encoding region of a large number of DRB1*0401 selected phage was sequenced. An alignment of the peptide encoding region of DR-bound phage, in comparison to the corresponding sequences from the unselected library, has revealed position"specific enrichments of certain amino acid residues, subsequently named class II anchors" (Figure 10.2a, b). As shown in Figure 10.3, the identified DRB1*0401 motifs consists of several anchor residues at anchor positions i, i+3 and i+5.

10.3 Conserved and Allele-Specific Anchor Residues Explain Promiscuity and Allele Specificity of HLA-DR/Peptide Interaction

Different groups have reported that certainMHC class II binding peptides are promiscuous in that they are capable of interacting with several, if not all, allelic DR molecules (Sinigaglia et al., 1988; Panina-Bordignon et al., 1989), whereas other peptides exhibit allele specific DR binding (Hammer et al., 1993). To understand the nature of promiscuous and allele specific peptide binding, we defined the peptide binding motifs of two additional MHC class II molecules, DRB1*0101 and DRB1*1101, using the phage technology and compared them to the DRB1*0401 motif (Figure 10.3). Each of the aligned sequence pools showed a striking enrichment (>80%) of aromatic residues near the NH_2-terminus of the peptide inserts (Hammer et al., 1993). We therefore named this position a "conserved anchor" and referred to it as position 1 (p1) anchor. A second conserved anchor position was found at relative position 4 where the aligned peptide pools showed an enrichment of hydrophobic residues. Further constraints were found at position 6. In the DRB1*0401 selected peptides we found most of the amino acids having hydroxyl groups (Thr and Ser). The positively charged amino acids Arg and Lys were enriched in the DRB1*0101 selected pool and small amino acids (Ala and Gly) were found for the DRB1*0101 binding peptides (Hammer et al., 1992). More extensive variability was found outside the anchor positions.

As Figure 10.3 shows, the anchors at position 6 clearly differ among these three DR alleles. This suggested that residues at this position confer allelic specificity to the binding. If correct, one would predict that the exchange of the residues at position 6 would modify the allele specificity of peptide binding. This was indeed confirmed by in vitro binding assays using single point mutations of

DRB1*0401

1	2	3	4	5	6	7	8	9
W			M		T			
Y			A		S			
			V					

DRB1*0101

1	2	3	4	5	6	7	8	9
Y			M		A			L
F			L		G			
W								

DRB1*1101

1	2	3	4	5	6	7	8	9
W			M		R			
			L		K			
			V					

Figure 10.3: Allele-specific MHC class II DR-peptide binding motifs.

both designer peptides and T cell epitopes (Hammer et al., 1993). The importance of conserved anchors for MHC/peptide binding was also demonstrated by such in vitro assays. The role of allele specific and conserved anchors in peptide binding to HLA-DR molecules, together with the observation that not all the anchors are used by individual ligands, appear to explain the molecular basis for promiscuity and allele specificity of peptide binding. The use of promiscuous or allele specific anchors in multiple combinations allows for a broad range of specificity from promiscuity to single allele specificity.

10.4 High-Stringency Screening and the Design of Short Peptide Antagonists

A striking characteristic of autoimmune diseases is the increased frequency of certain HLA class II alleles among affected individuals. Although the pathogenesis of autoimmune diseases is not known, one possibility is that disease-associated HLA class II molecules have the capacity to bind and present autoantigenic peptides to T cells. These peptides could activate autoreactive $CD4^+$ T cells,

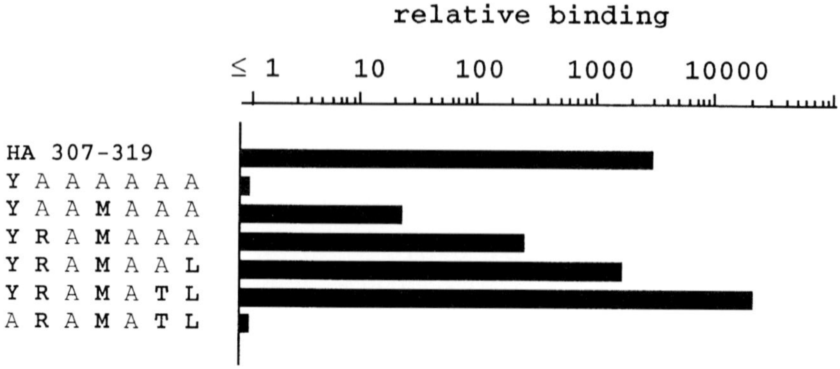

Figure 10.4: Construction of high-affinity-binding short peptides by anchor addition. Binding of the high-affinity natural epitope HA 307–319 is included for comparison.

which in turn induce a cascade of inflammatory lymphokines leading to chronic inflammation. Blocking the antigen-presenting capacity of disease-associated HLA class II molecules could represent a suitable target to prevent and possibly also treat autoimmune diseases (Adorini et al., 1990).

The knowledge of anchor positions and other general rules for peptide-binding toMHC molecules should prove very useful for the design of MHC-specific antagonists. Although this strategy should be ultimately directed at the development of non-peptide antagonists that block the peptide binding site of disease-linked class IIMHC molecules, the identification of short, high affinity peptide antagonists would provide very useful leads for a rational drug design approach.

To identify short peptide antagonists, we increased the stringency of the phage library screening to identify possible additional anchors for the disease-associated DRB1*0401 allele. An additional pH3 preelution permitted us to introduce more stringency into the screening by selecting only the phage eluted at pH2.2 subsequent to the pH3 preelution step. With this new approach, most of the selected phage sequences contained aromatic residues at p1 and Thr or Ser at p6, i.e., the expected residues at the two dominant anchor sites (Figure 10.2), half of the sequences contained hydrophobic residues (Met, Ala) at p4. However, additional enrichments could also be seen: half of the independent sequences had Arg at p2, small amino acids Gly and Ala at p3, and Leu at p7. Taken together, amino acid preferences could be demonstrated at six out of seven positions. Most interestingly, the incorporation of an increasing number of these anchors into short designer peptides (hexamers or heptamers) resulted in a gradual increase of binding affinity to the level of 13-residues-long high affinity epitopes (Figure 10.4). These results demonstrate the value of phage displayed peptide repertoires for the identification of MHC-specific antagonists.

10.5 Anchor Residues Interact with Pockets of the MHC Class II Peptide Binding Cleft

Many of the above mentioned characteristics of HLA-DR bound peptides are in accordance with the recent x-ray crystallographic structural analyses of the class II DRB1*0101 molecule and the DRB1*0101/HA 307–319 MHC-peptide complex.

X-ray crystallographic studies indicated that different peptides bind with a similar conformation to HLA-DR molecules (Brown et al., 1993). This observation is supported by the analysis of the large number of peptide sequences selected from M13 phage display libraries. Although these peptides differ in their primary sequence, most of them share perfectly spaced anchor residues, suggesting an overall similar and sequence independent peptide conformation. Recent structural data from Stern et al. (Stern et al., 1994) indicated that conserved hydrogen bonds between HLA-DR residues and the peptide main chain may be responsible for the similar conformation of class II bound peptides.

The anchor closest to the N-terminus of class II bound peptides is defined as the p1 anchor (see above). The p1 anchor seems to be prominent in DR molecules, in that an Ala-substitution at this position abrogates peptide binding completely, whereas the elimination of other anchors results only in partial loss of binding affinity (Figure 10.4). The p1 anchor is conserved, since each of the DR-selected peptide pools show a striking enrichment of aromatic residues at this position (Figure 10.3). These results fit well with the structure of HLA-DRB1*0101 molecules. The cleft contains only one deep pocket lined with non polar residues capable of accommodating the dominant hydrophobic anchor residues. The deep pocket is built by the invariant α-chain and by a fairly conserved part of the β-chain, explaining the conserved nature of the p1 anchor position.

X-ray analysis also revealed that most of the polymorphic residues of the HLA-DR β-chain are located within the peptide binding site of the class II molecule. Clusters of polymorphic residues shape shallow pockets which seem to interact with allele specific anchor residues, e.g. explaining the allele specific nature of the p6 anchor position.

10.6 Refinement of Peptide Motifs and Prediction of MHC Class II/Peptide Interaction

Methods to predict regions in protein sequences capable of binding human MHC class II proteins would be very valuable for many immunological applications, e.g. for the design of subunit vaccines or the identification of potential autoantigenic peptides. Considering the synergistic nature and position specificity of

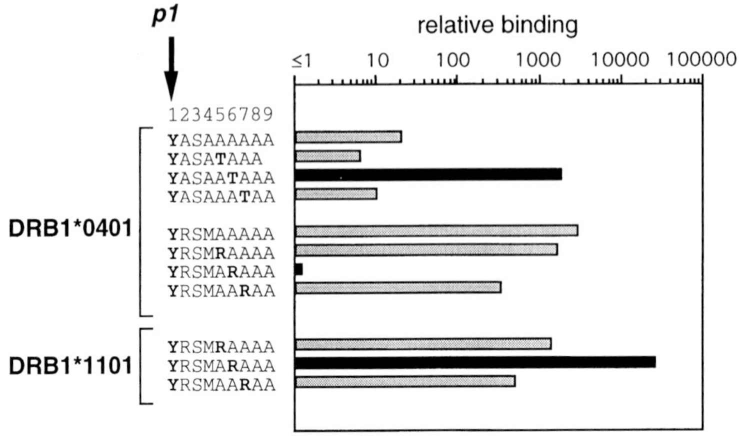

Figure 10.5: Effects of anchor shifting and inhibitory residue shifting on peptide binding to three allelic DR molecules.

class II anchors (Figure 10.4, 10.5), it is reasonable to assume that peptides with several anchors in frame are more likely to bind to a particular class II molecule than peptides lacking most or all anchors in frame. We have investigated this hypothesis by selecting a set of sequences from different proteins containing 3 to 4 anchor residues in frame for DRB1*0101, and another set without anchors in frame. Most of the peptides with anchors bound to DRB1*0101 (Hammer et al., 1994a). In contrast, all peptides without anchors in frame failed to bind, supporting the identified rules of MHC-peptide interaction. However, many peptides with two anchors in frame failed to bind to DR molecules, despite the fact that known T cell epitopes and natural eluted peptides often contain only two anchor residues (Hammer et al., 1993). This suggested that additional rules remain to be identified for a more precise, quantitative prediction of MHC binding sites.

Mutational analyses with T cell epitopes and designer peptides revealed negative effects of particular amino acids on MHC-peptide binding (Hammer et al., 1994a; Sette et al., 1993). Subsequent experiments demonstrated position and allele specific properties of these inhibitory residues (Figure 10.5). Since a quantitative prediction of MHC binding sites requires the understanding of the effects of both anchor and inhibitory residues, we developed a new method for predicting class II MHC binding peptides. The method is based on the hypothesis that (i) most peptides bind with similar conformation to DR molecules (see above), (ii) that peptide binding correlates with the net result of all side chain effects in any given peptide, and that (iii) most side chain effects depend on their relative position within the p1-peptide-frame, rather than on the remaining peptide sequence (Hammer et al., 1994b).

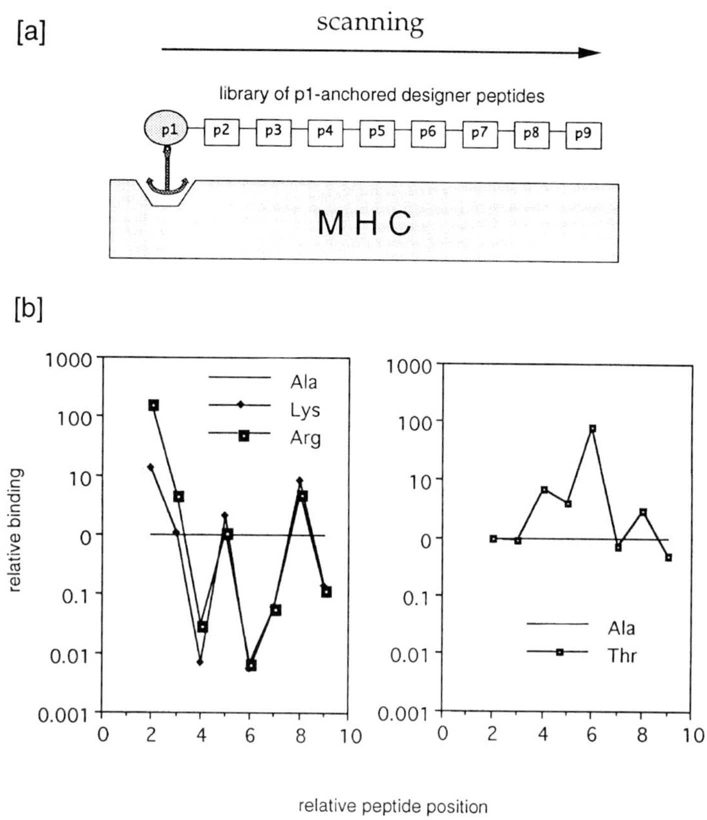

Figure 10.6: The principle of side chain scanning of p1-anchored designer peptide libraries.
a. Schematic representation of the anchoring of the peptide library.
b. Example: scanning of DRB1*0401 with Arg, Lys and Thr and the identification of position specific anchor and inhibitory residues.

A quantitative analysis of the side chain effects at each peptide position made it necessary to use synthetic peptide libraries. We used libraries of short, p1-anchored and Ala-based designer peptides where all naturally occurring L-amino acids had been substituted at each position from 2 to 9 (Figure 10.6). Nine residue-long peptides cover most of the potential side chain interactions within the MHC cleft, and show reduced main chain interaction compared to longer peptides (Hammer et al., 1994a), therefore amplifying the effect of possible anchor and inhibitory residues. P1-anchor residues are needed for high affinity binding, even if peptides contain all additional anchor residues. Thus, the usage of both the p1-anchoring and short peptides highly decreased the probability of shifts within the p1-frame.

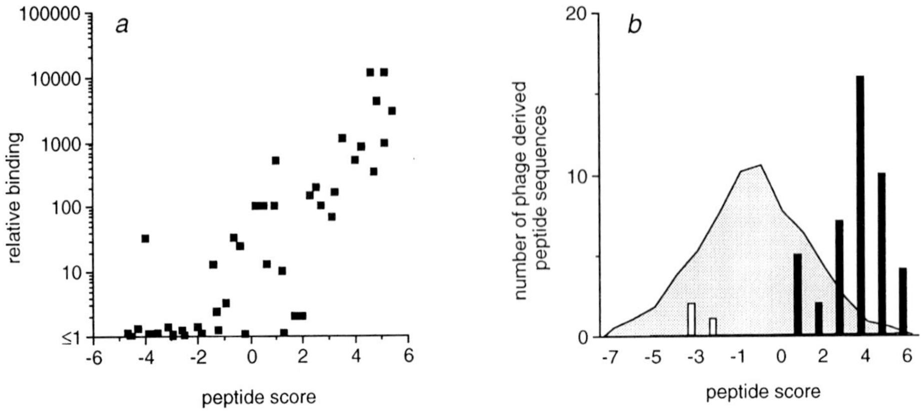

Figure 10.7: Correlation between peptide scores and peptide affinity.
a. The peptide "scores" of randomly selected peptides (nonamers) correlates with their affinity to DRB1*0401.
b. The peptides of DRB1*0401-selected bacteriophage (black bars) show scores corresponding to high affinity binding. Only three peptides of the DRB1*0401-selected phage pool had low scores (white bars). None of them bound to the DRB1*0401 molecule. The grey surface indicates the average distribution of peptide scores in human proteins.

We scanned the p2 to p9 positions for the effect of each amino acid side chain on binding to the DRB1*0401 molecule. The IC_{50} data derived from side chain scanning of p1-anchored designer libraries were processed into a software predicting MHC-binding regions in proteins. Protein sequences are first scanned for p1 anchor residues. After having identified a p1 anchor, the program locates the amino acid residues at relative positions 2 to 9. Next, values obtained from the side chain scanning of p1-anchored peptide libraries are assigned to each of the amino acid residues of the selected protein region. The sum of these values gives a score indicating the predicted peptide binding affinity (Hammer et al., 1994b). To validate our scoring system we have used this algorithm to analyse an unbiased set of peptides. As shown in Figure 10.7 there is a very good correlation between binding affinities and scores. The peptides that apparently have the highest affinities are also the ones that have the highest scores. This indicates that this algorithm should be an efficient means of identifying peptide binding to a given DR allele, and should help in the identification of T cell epitopes in natural polypeptides.

10.7 Changing the Fine Specificity of a Class II MHC Pocket

Susceptibility to rheumatoid arthritis (RA) is specifically associated with class II MHC alleles DRB1*0401, DRB1*0404, and DRB1*0101 (Nepom and Erlich, 1991). Interestingly, the DRβ chains encoded by these genes possess a "shared epitope" formed by a short stretch of amino acids (at positions 67 to 74) that is highly conserved among RA-associated molecules (Gregersen et al., 1987). Since DRB1*0402, a closely related molecule not associated with RA, differs from the RA-linked DRB1*0401 and DRB1*0404 molecules only in the shared epitope region, this part of the molecule is likely to be critical for disease association.

Crystallographic studies and MHC modelling indicated that MHC position β-71 forms part of the p4 pocket. In an attempt to gain insight into the mechanism of RA association, we have studied the influence of position 71 on peptide binding specificity. The p4 anchor appeared to be conserved in peptides selected by DRB1*0101, DRB1*0401 and DRB1*1101 (Figure 10.3) because Met was the preferred residue. However, the use of p1-anchored designer peptide libraries revealed striking differences in the effect of charged residues at p4. Namely, negative charged residues were accepted and positive charged residues not by the RA-associated DRB1*0401 subtype, and the opposite was the case for the non-associated DRB1*0402 subtype (Hammer et al., 1995). We could demonstrate that a site directed mutant molecule that differs from DRB1*0401 only in a single Lys to Glu exchange at position 71 of the β-chain exhibits a binding specificity similar to that of DRB1*0402 (Figure 10.8). Thus, the exchange of a singe amino acid residue at position 71 seems to account for most major differences in binding specificity between the RA-associated and non-associated DR4 allotypes.

10.8 Peptide Libraries and MHC: An Outlook

Each MHC class II molecule binds to a large natural repertoire of peptides. We have demonstrated that the screening of artificial peptide repertoires displayed on the M13 phage surface allows for the identification of general rules governing MHC class II-peptide interaction, e.g. the identification of anchor residues or exactly spaced anchor positions in class II bound peptides. We also indicated the limits of an approach which is solely based on screening large molecular repertoires. The identification of detailed binding motifs is not possible because of the difficulty to identify inhibitory residues. Therefore, a refinement of motifs is necessary for applications such as epitope prediction or the analysis of the effect of single MHC residues on peptide binding. The combination of both large molecular peptide repertoires and small libraries of single substituted designer peptides was shown to be sufficient for the identification of accurate peptide binding motifs.

More than 50 different human MHC class II molecules have been identified so far. The challenge for the future is the identification of peptide motifs for all class II molecules. This will allow to predict pathogen-derived promiscuous peptides capable of binding to most class II MHC molecules in a given human population, thus facilitating the development of subunit vaccines. In addition, a large effort in the identification of class II binding motifs will unquestionably provide valuable information to important biological fields such as immune recognition and autoimmunity.

[a]

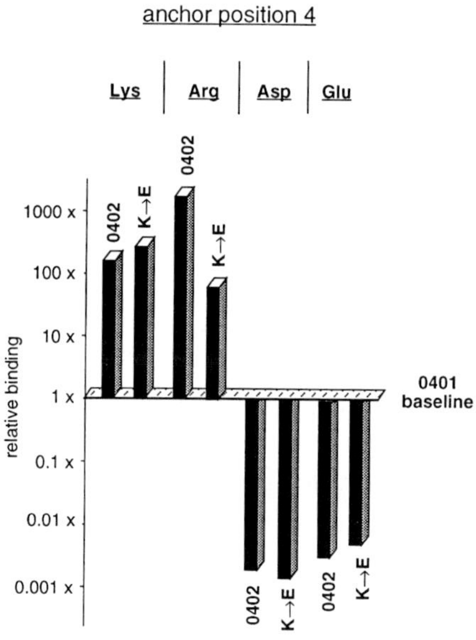

[b]

HLA-DR	DRβ residues at position			
	67	70	71	86
0401	L	Q	K	G
0402	I	D	E	V
K→E	-	-	E	-

Figure 10.8: The effect of position 71 on the peptide binding specificity of RA-associated and non-associated DR4 subtypes.
a. The effect of charged residues is shown for the anchor position 4. Relative binding data were plotted in comparison to binding data of DRB1*0401 as baseline
b. Amino acid differences of the shared epitope region between the DR molecules used in (a).

References

Adorini, L., Barnaba, V., Bona, C., Celada, F., Lanzavecchia, A., Sercarz, E., Suciu-Foca, N. and Wekerle, H. (1990) New perspectives on immunointervention in autoimmune diseases, Immunol. Today *11*, 383–386.
Brown, J.H., Jardetzky, T.S., Gorga, J.C., Stern, L.J., Urban, R.G., Strominger, J.L. and Wiley, D.C. (1993) 3-Dimensional structure of the human class-II histocompatibility antigen HLA DR1, Nature *364*, 33–39.
Cwirla, S.E., Peters, E.A., Barret, R.W. and Dower, W.J. (1990) Peptides on phage: A vast library of peptides for identifying ligands, Biochemistry *87*, 6378–6382.
Devlin, J.J., Paniganiban, L.C. and Devlin, P.E. (1990) Random Peptide Libraries: A Source of Specific Protein Binding Molecules, Science *249*, 404–406.
Gregersen, P.K., Silver, J. and Winchester, R.J. (1987) The shared epitope hypothesis: an approach to understanding the molecular genetics of susceptibility to rheumatoid arthritis, Arthritis Rheum. *30*, 1205.
Hammer, J. and Sinigaglia, F. (1995) Techniques to identify the rules governing class II MHC-peptide interaction, In MHC: A Practical Approach (Butcher, G. and Fernandez, N.; eds.) Oxford University Press, in press.
Hammer, J., Valsasnini, P., Tolba, K., Bolin, D., Higelin, J., Takacs, B. and Sinigaglia, F. (1993) Promiscuous and allele specific anchors in HLA-DR-binding peptides, Cell *74*, 197–203.
Hammer, J., Takacs, B. and Sinigaglia, F. (1992) Identification of a motif for HLA-DR1 binding peptides using M13 display libraries, J. Exp .Med. *176*, 1007–1013.
Hammer, J., Belunis, C., Bolin, D., Papadopulos, J., Walsky, R., Higelin, J., F., S. and Nagy, Z. (1994a) High affinity binding of short peptides to MHC class II molecules by anchor combinations, Proc. Natl. Acad. Sci. USA. *91*, 4456–4460.
Hammer, J., Bono, E., Gallazzi, F., Belunis, C., Nagy, Z. and Sinigaglia, F. (1994b) Precise prediction of MHC class II-peptide interaction based on peptide side chain scanning, J. Exp. Med., *180*, 2353–2358.
Hammer, J., Gallazzi, F., Bono, E., Karr, R.W., Guenot, J., Valsasnini, P., Nagy, Z. and Sinigaglia, F. (1995) Peptide binding specificity of HLA-DR4 molecules: correlation with rheumatoid arthritis association, J. Exp. Med. *181*, 1847–1855.
Nepom, G.T. and Erlich, H.A. (1991) MHC class II molecules and autoimmunity, Annu. Rev. Immunol. *9*, 493–520.
Panina-Bordignon, P., Tan, A., Termijtelen, A., Demotz, S., Corradin, G. and Lanzavecchia, A. (1989) Universally immunogenic T cell epitopes: promiscuous binding to human MHC class II and promiscuous recognition by T cells, Eur. J. Immunol. *19*, 2237–2242.
Scott, J.K. and Smith, G.P. (1990) Searching for Peptide Ligands with an Epitope Library, Science *248*, 386–390.
Sette, A., Sidney, J., Oseroff, C., Delguercio, M.F., Southwood, S., Arrhenius, T., Powell, M.F., Colon, S.M., Gaeta, F.C.A. and Grey, H.M. (1993) HLA DR4w4-binding motifs illustrate the biochemical basis of degeneracy and specificity in peptide DRinteractions, J. Immunol. *151*, 3163–3170.
Sinigaglia, F., Guttinger, M., Kilgus, J., Doran, D.M., Matile, H., Etlinger, H., Trzeciak, A., Gillesse, D. and Pink, J.R.L. (1988) A malaria T-cell epitope recognized in association with most mouse and human MHC class II molecules, Nature *336*, 778–780.
Stern, L.J., Brown, J.H., Jardetzky, T.S., Gorga, J.C., Urban, R.G., Strominger, J.L. and Wiley, D.C. (1994) Crystal structure of the human class II MHC protein HLA-DR1 complexed with an influenza virus peptide, Nature *368*, 215–221

D Phage Display of Protein Domains

11 Isolating High Affinity Human Antibodies from Phage Repertoires

Kevin FitzGerald, David Chiswell, John Earnshaw, Rodger Smith, John Kenten, Richard Williams and John McCafferty

11.1 Introduction

Recent advances in the isolation of high affinity human antibodies owe much to lessons learnt from the natural immune system. This article describes the key advances and illustrates the practical process with a description of the isolation of antibodies directed against a small organic hapten.

The immune system creates a large and highly diverse repertoire of B-cells that mirrors the diversity of potential antigenic insult. The diversity of this repertoire is achieved during the ontogony of the B-cell using a variety of strategies including an essentially random combinatorial assortment of many antibody gene segments taken from a germline pool. Each mature B cell thus contains the rearranged genetic information for a single antibody "species" which it displays on its surface (Tonegawa 1983).

The specific interaction of antigen with a particular membrane bound antibody acts as a trigger for the growth and differentiation of that B-cell. The result is an expanded clone of short lived antibody secreting plasma cells and a long lived memory cell pool which accumulates mutations within the antibody variable region genes. The reappearance of antigen triggers these memory cells to differentiate into plasma cells. There is a bias towards the triggering of cells that display antibodies which can forge higher affinity interactions with the antigen and this selective enrichment for more effective binders results in an overall improvement in the quality of the response; a process known as affinity maturation (Berek and Milstein, 1988).

The development of techniques to enrich for the presence of B cells specific for a particular immunogen and the immortalisation of these cells via fusion to immortal partners has led to the isolation of countless numbers of rodent monoclonal antibodies (Kohler and Milstein, 1975). Similar success in obtaining human antibodies has been elusive. Ethical considerations that prevent the hyperimmunisation of humans or the removal of tissues rich in antibody-secreting B-cells from naturally immune individuals have hampered the development of techniques to isolate human monoclonal antibodies. In vitro immunisation techniques have not been able to overcome these problems. Further, it has proved difficult to immortalise human B-cells from peripheral blood; EBV transformation or the use of mouse myelomas as fusion partners typically results in the preferential loss of human chromosomes and consequently unstable cell lines producing poor yields of low affinity antibody (Winter and Milstein, 1991; James and Bell, 1987).

While non-human antibodies have undergone trials for clinical administration in humans their utility has commonly been limited due to anti-immunoglobulin responses (for example HAMA – Human Anti Mouse Antibody – responses) in the recipient. The expectation that the therapeutic potential of antibodies would only be fully realised if human proteins could be readily isolated prompted the effort to modify existing rodent antibodies using recombinant DNA technologies such that they resembled, as far as was possible, human proteins (Neuberger et al., 1985; Jones et al., 1986; Riechmann et al., 1988; Verhoeyen and Riechmann, 1988; Queen et al., 1989). However, several key developments have combined to make it now possible to isolate fully human antibodies with specificities to any desired target.

Techniques were developed to express active antibody fragments in bacteria via export to the periplasm (Skerra and Pluckthun, 1988; Better et al., 1988). The demonstration that antibody genes could be amplified by means of the polymerase chain reaction (Saiki et al., 1985) using consensus primers that each recognised many different antibody genes (Orlandi et al., 1989) permitted the combinatorial cloning of large repertoires of human (or rodent etc.) VH and VL genes in the correct format for expression (Ward et al., 1989; Sastry et al., 1989; Huse et al., 1989; Marks et al., 1991). This ability to immortalise the antibody genes has circumvented the problems inherent in immortalising the B-cells that carry them. Another major advance was the development of a powerful method for selecting the rare clones that have desired properties from repertoires that can contain millions (or even billions) of members. Antibody fragments are cloned such that the are expressed fused to one of the capsid proteins of the filamentous E.coli bacteriophage fd. The minor protein coded by gene III (pIII) is found in 3–5 copies at one end of the virion (for a structural review on filamentous E.coli bacteriophages see Rasched and Oberer, 1986) and is most commonly the chosen site for phage display (Smith, 1985). The antibody fragments so expressed fold correctly to form functional protein displayed on the surface of assembled virions (McCafferty et al., 1990). Repertoires of phage antibodies can be incubated with an

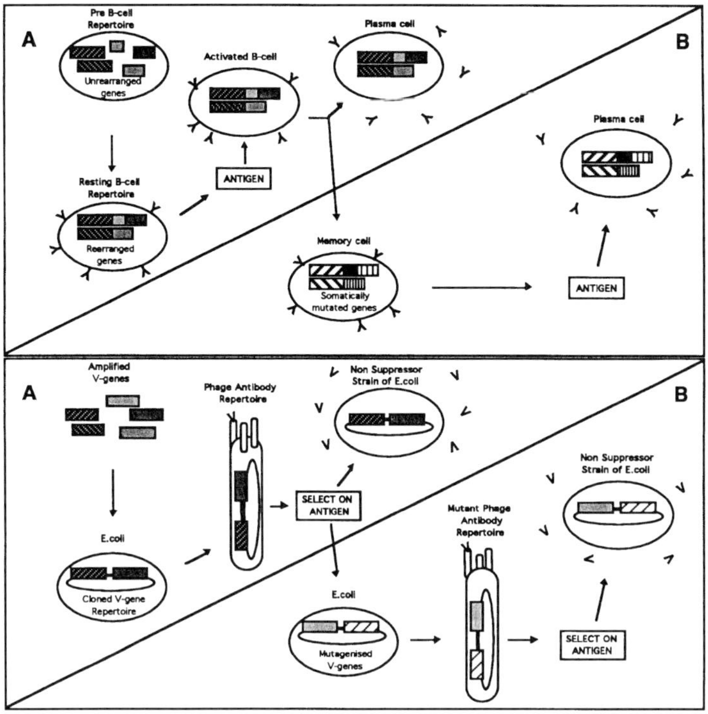

Figure 11.1: A. Antibody production by the phage display system (lower panel) mimics the immune system (top panel). In both systems a combinatorial rearrangement of antibody gene segments results int the generation of a diverse repertoire of antibody molecules for surface display. Antigen stimulated B-cells differentiate into antibody secreting plasma cells and antigen selected phage can be infected into a non suppressor strain of *E. coli* for soluble expression of antibody fragments.
B. Affinity maturation by the phage display system mimics the immune system. Activated B-cells differentiate into memory cells which are subjected to somatic mutation. Selected phage can be infected into a suppressor strain of *E. coli* for mutagenesis of the V-genes. High affinity antibodies arise from re-stimulation of mutated memory cells or re-selection of mutated phage.

immobilised target antigen, non-binders washed away and antigen specific clones eluted, infected into E.coli and grown for subsequent rounds of selection (Clackson et al., 1991; Marks et al., 1991; Barbas et al., 1991). The power of the system resides in the fact that both protein and gene (packaged within the virion) are co-selected and, in this respect, the phage thus mimics the B-cell (Figure 1).

Antibody fragments can be cloned for surface display within the phage genome (McCafferty et al., 1990) or alternatively within a phagemid vector (Hoogenboom et al., 1991; Barbas et al., 1991). Phagemids are plasmid vectors that posses the M13 origin of replication to allow packaging within nascent M13 (or Fd) virions. The remaining viral proteins that are required for virion production are provided by a helper virus which has a defective origin so that its genome does not compete favourably with the phagemid DNA for packaging (Yanisch et al., 1985; Vieira and Messing, 1987). With phagemid vectors, the competition for display that exists between antibody-pIII fusions and helper derived wild type pIII combines with proteolysis of displayed antibody-pIII fusions to result in an average of significantly less than one antibody molecule being displayed per phage particle. It is generally held that phage vectors, by contrast, are multivalent since all gene III products would be fused to antibody sequences (Lowman et al., 1991; Garrard et al., 1991) however there is evidence that in many cases proteolysis also results in an average of less than one antibody molecule per phage albeit with a higher fraction of phage carrying intact fusions compared to phagemid derived particles (unpublished data). The principle benefit in using phagemid vectors resides in the increased efficiency of transformation into E.coli that is afforded which thus makes possible the generation of larger repertoires. Truly multivalent phage can be generated by fusing the antibody fragments to the multicopy capsid protein encoded by gene VIII (pVIII) (Kang et al., 1991; Huse 1991; Chang et al., 1991; Huse et al., 1992). Selections using this type of vector system are likely to be poorly discriminating on the basis of affinity since even low affinity antibodies will bind tightly to immobilised antigens due to avidity effects (an apparent high affinity interaction arising from the cooperation between two or more antigen binding sites).

Repertoires have been made using antibody genes derived from immunised rodents (Clackson et al., 1991), naturally immune humans (Barbas et al., 1991) and from combinations of natural antibody sequences and synthetic sequences (for example, repertoires of random VH CDR3 sequences have been synthetically incorporated into a set of human germline VH genes (Hoogenboom and Winter, 1992) or into a single human antibody gene (Barbas et al., 1992a). To obtain human antibodies to target antigens for which an immune source is not available, for example to toxins or self antigens, non-immunised phage libraries must be used (Marks et al., 1991). Human antibodies to a range of foreign antigens (BSA, TEL, phOx, bovine thyroglobulin) and self antigens (human TNFα, human thyroglobulin, CEA, monoclonal antibodies, mucin and blood groups) have been isolated from a single non-immune repertoire (Marks et al., 1991; Griffiths et al., 1993; Marks et al., 1993). This highlights the flexibility of the system in that a single library can serve as a rich resource eliminating the need to make new repertoires for each target antigen. As with the in vivo antibody response however, clones that are isolated from such repertoires tend to have only moderate affinities (typically 0.1–1 μM). For most applications higher binding affinity is required and consequently primary isolates require modification to improve their affinities and enhance their utility. A great strength of the phage system is that it allows the affinity maturation process to be mimicked in vitro.

There are a number of strategies available to mutate primary isolates thereby creating new second generation repertoires (Clackson et al., 1991; Kang et al.,

1991; Marks et al., 1992). Selections of these libraries and isolation of improved binders provides the raw material for subsequent mutation and selection. The generation of hierarchies of phage libraries can be accompanied by high stringency selections (Hawkins et al., 1992) and in most cases leads to the isolation of human antibodies with affinities comparable to those obtained from immunised rodent sources.

We illustrate in this article the isolation, without immunisation, of high affinity human antibody fragments to a transition state analogue hapten via the selection of a large repertoire of phage antibodies and the in vitro affinity maturation of primary isolates.

11.2 High Affinity Human Antibodies to RT3

11.2.1 Primary Isolates

The transition state analogue RT3 is a small organic hapten consisting of two aromatic groups connected by a phosphonate group and was synthesised to derive potentially catalytic antibodies to the corresponding ester (McCafferty et al., 1994). A single chain Fv (scFv) phagemid repertoire containing 2.9×10^7 clones derived from the peripheral blood lymphocytes of a non immunised human donor, was used to select human binders to RT3: a scFv consists of a VH domain and VL domain covalently joined by the 15 amino acid flexible peptide linker (Gly4-Ser)3 (Huston et al., 1988). RT3 coupled to BSA was coated onto plastic tubes and the immobilised hapten was used to select binders from the rescued phage antibody repertoire. Non-specific phage were removed by washing and RT3 binders were eluted, infected into E.coli and plated onto selective medium. In all, the process of rescue and selection was repeated four times (for protocol details see Jackson et al., 1992).

An amber mutation between the gene III nucleotide sequence and the antibody sequence permits the facile switching between phage displayed antibody and soluble antibody production by simply infecting the phage into either suppressor or non suppressor strain of E.coli respectively (Hoogenboom et al., 1991). Hence, after the third and fourth rounds of panning eluted phage were infected into the non-suppresser strain HB2151 and soluble scFv induced from individual colonies. Binding to RT3-BSA was detected in ELISA by means of the monoclonal antibody 9E10 that binds to a peptide tag, a fragment of the c-myc oncogene, which was fused to the C-terminus of the scFvs (Munro and Pelham, 1986). The proportion of binders in the eluted population rose from 46% after three rounds of panning to 83% after four rounds. This increase in the number of specific phage is a useful indication that the panning process has been successful. "Fingerprints" of individual clones were obtained by PCR amplification of the scFv construct followed by digestion with the common cutting restriction enzyme BstN1 (Gussow and Clackson, 1989). The clones could be placed into three groups on the basis of digestion pattern and these PCR groups were also found to coincide with strength of ELISA signals achieved by induced supernatants. Subsequent sequencing (Sanger et al., 1977) confirmed that the three groups comprised multiple isolates of three different clones (designated RT3:1, RT3:47 and RT3:61). Specificity to RT3 was confirmed by assaying binding to BSA and by competition with soluble RT3 (data not shown).

11.2.2 Chain Shuffling

The first route taken to mutate each of the 3 RT3-specific clones was to fix one of the V-domains and combining it with a population of partner chains taken from the unselected repertoire. This strategy was adopted to enable the isolation of clones with more optimal VH-VL pairings. Since no information was available as to which V-domain was most important for RT3 binding, the VL and VH chains from each clone were "shuffled" separately and the resulting populations pooled. The 3 new libraries were then subjected to two rounds of panning against RT3-BSA. To monitor the progress of the selection process polyclonal phage derived from each panning were prepared and screened for binding to RT3. The ELISA signals produced by equal titres of phage were elevated for the pan-2 phage compared to the pan-1 phage signalling the anticipated enrichment of RT3-specific phage (data not shown).

Individual colonies were grown from the populations derived from each round of panning and soluble scFv was induced and screened for binding to RT3-BSA. The RT3:1 chain shuffled library resulted in 88% positive clones after one round of panning and 95% after two rounds. The RT3:47 chain shuffled library resulted in 66% positive clones after one round of panning and 73% after two rounds. The RT3:61 chain shuffled library was only screened after two rounds of panning and 44% bound to RT3-BSA. The chain shuffled populations required fewer rounds of selection (1–2) than the unselected repertoire (3–4) in order to isolate binders. This illustrates the expected richness of the chain shuffled libraries for RT3 binders compared to the unselected repertoire.

The sequence diversity of the chain shuffled isolates was estimated by sequencing 15 clones. All had retained the original heavy chain suggesting that the VH domain was more crucial for forming an RT3-specific binding site than the VL domain. There was a range of diversity in the new light chain partners from single residue changes to completely different VL sequences. No two sequences were identical which suggested that there were few (if any) duplicates among the 187 RT3 binders that had been isolated.

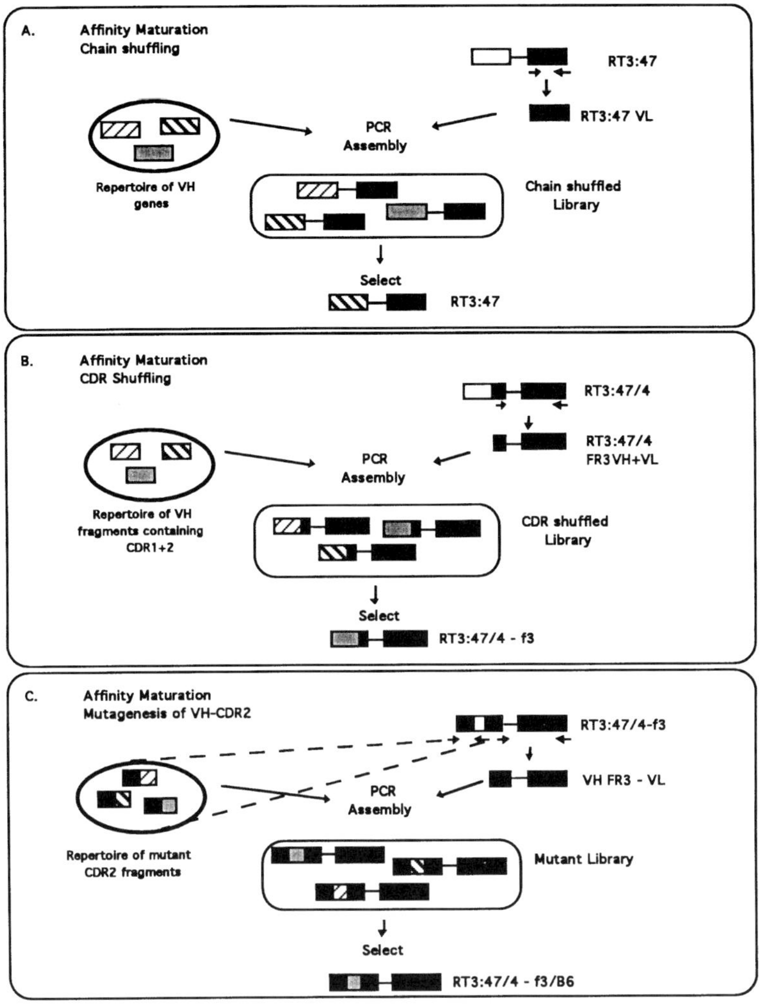

Figure 11.2: Affinity maturation of RT3:47. A: A repertoire of VH fragments was combined with the VL fragment of RT3:47. The chain shuffled library was selected against antigen and the improved RT 3 binder RT3:47/4 was isolated. B: A repertoire of VH subfragments containing the CDR1 and CDR2 regions was combined with the remaining portion of RT3:47/4. The CDR shuffled library was selected against antigen and the improved RT3 binder RT3:47/4-f3 was isolated. C: A repertoire of VH subfragments with 5 residue positions in the CDR2 region randomised was combined with the remaining portion of RT3:47/4-f3. The mutant library was selected against antigen and the improved RT3 binder RT3:47/4-f3/B6 was isolated.

Competitive ELISA experiments with soluble hapten were used to rank the clones and one of the isolates from the RT3:47 chain shuffled library was identified as probably the best binder (equilibrium dissociation constant measured at 105 x 10^{-9}M). This clone, designated RT3:47/4, was used as the template for the next step in the affinity maturation process, CDR shuffling.

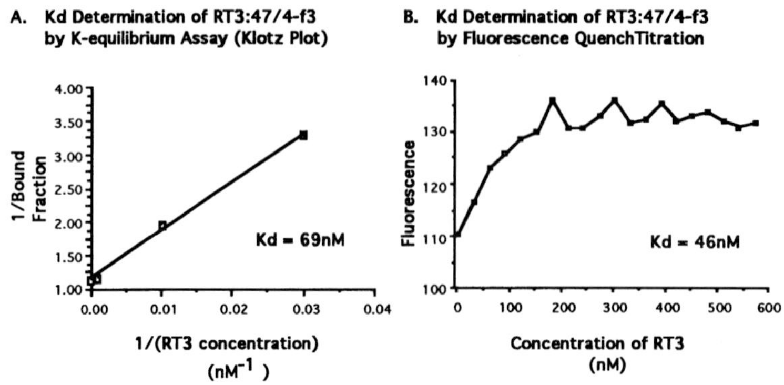

Figure 11.3: The equilibrium dissociation constant (Kd) for RT3:47/4-f3 was measured by K-equilibrium assay (A) and by fluorescence quench titration (B).

11.2.3 CDR Shuffling

A repertoire of VH sub-fragments containing the CDR1 and CDR2 regions was amplified using PCR and assembled with a fragment of RT3:47/4 which contained the remaining portion of the VH domain plus the entire VL domain. This resulted in the generation of a library in which the CDR1 and CDR2 regions (plus surrounding framework regions) of RT3:47/4 were effectively randomised (Figure 2.2). In a parallel experiment the CDR1+2 region of RT3:47/4 and the whole of the light chain were fixed and assembled with a repertoire of amplified CDR3 fragments.

Both libraries were panned once and individual colonies screened for binding to RT3-BSA. From the selected CDR1+2 shuffled library 95 binders were identified out of 376 and from the panned CDR3 shuffled library 118 binders were identified out of 376 clones screened. The nucleotide sequences of 60 clones were established and were found to be closely related (many differing by a small number of point mutations only) but no duplicate sequences were found. This suggested that there was a high level of diversity among the 213 identified binders.

The affinities of the CDR shuffled binders that had been sequenced were measured by the K-equilibrium ELISA method of Friguet et al (Friguet et al., 1985) in conjunction with the data transformation method described by Klotz (Klotz 1953). The best clone identified was from the CDR1+2 shuffled population and designated RT3:47/4-f3. This had an apparent Kd of 69×10^{-9}M. The Kd of this clone was also measured by fluorescence quench titration (Foote and Milstein, 1991) using material prepared by immobilised metal affinity chromatography (IMAC), which utilises a hexa-histidine tag which is fused to the C-terminus of the protein (McCafferty et al., 1994), followed by size exclusion chromatography (FPLC) to obtain monomeric scFv. The addition of soluble RT3 resulted in an enhanced fluorescence rather than the more typical reduction in fluorescence but the overall fluorescence change still allowed the quantitation of the affinity to be made. The value obtained was 46×10^{-9}M. The close agreement between the two methods of affinity measurement suggested that the K-equilibrium assay was probably giving an accurate ranking of the isolated binders (Figure 11.3).

11.2.4 Directed Mutagenesis

By comparing the sequences and affinity ranking of the CDR1+2 shuffled clones it was possible to identify a group of 5 residues in CDR2 of the VH that appeared to influence, both in a positive and negative way, the interaction between antibody and antigen. These residues were therefore targeted for further diversification with the hope that subsequent selection would yield a clone in which optimal residues would occur at all 5 positions. A mutagenic oligonucleotide was synthesised which spanned the CDR2 region of RT3:47/4-f3 to randomise these key residues (Figure 11.4). The mutagenic oligonucleotide as used to amplify a segment of DNA stretching from CDR2 of the VH to the 5′ flanking region of the VH gene. An overlapping reverse oligonucleotide was synthesised to amplify a

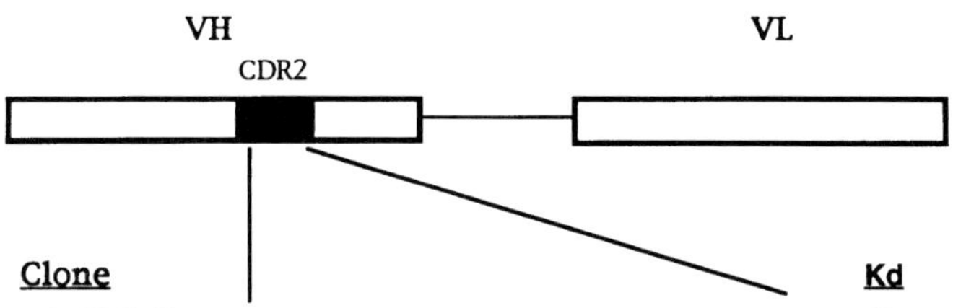

Clone											Kd
RT3:47/4-f3	S	I	S	S	S	S	G	D	T	Y	69nM
12	-	-	-	T	-	G	S	S	-	-	100nM
19	-	-	-	G	-	G	-	T	I	-	133nM
21	-	-	-	-	T	G	S	S	-	-	137nM
23	-	-	-	G	T	G	-	S	A	-	315nM
Mutant Oligo:	X	I	S	X	S	X	S	X	X	Y	

Figure 11.4: Directed mutagenesis. By relating sequences of CDR shuffled clones to their affinity for RT3 (as measured by K equilibrium assay, e.g. above, or by IC_{50}, not shown) 5 residues in the CDR2 region of the VH domain were identified that appeared to influence the interaction with antigen. A mutagenic oligonucleotide was designed that randomised the CDR2 VH sequence of RT3:47/4 at the identified residue positions (represented by an X). A fragment of RT3:47/4 amplified using the mutagenic oligo and a 5′ flanking primer was PCR assembled with a fragment of RT3:47/4 amplified using a primer that contained reverse complementary sequence to the mutant oligo and a 3′flanking primer. The resulting mutant repertoire was selected against antigen using high stringency conditions and the RT3 binding clone RT3:47/4-f3/B6 was identified which had a Kd for RT3 of 30×10^{-9} M.

fragment from CDR2 of the VH to the 3′ flanking region of the VL. The amplification products were assembled by PCR to derive a repertoire of scFv fragments containing the randomised CDR2 VH sequence. These were cloned into a phagemid vector for surface display and selection against RT3 (Figure 11.2, 11.3).

11.2.5 Stringent Selections

Previously, selections had been carried out with the generation of a diverse set of RT3 binders as the major goal. The selection strategy for the CDR2 VH mutant library was altered to bias the selection in favour of high affinity binders and the use of tubes coated with RT3-BSA was thus replaced with soluble biotinylated RT3-BSA. With this selection system the phage antibody library is added to the biotinylated antigen and bound phage are captured on streptavidin coated magnetic beads. Non binders are removed by washing the beads and specific phage eluted in the normal way (Hawkins et al., 1992).

Biotinylated selections permit greater control of selection conditions and, in particular, by setting the concentration of available antigen at a value below a desirable Kd, high affinity binders can be preferentially isolated (Hawkins et al., 1992). The mutant library was thus selected twice against biotinylated RT3-BSA at a concentration of 0.1 nM. 188 clones were screened and 150 were found to be positive for RT3.

The affinities of RT3 binding clones were measured by K equilibrium ELISA and several clones were identified that had better affinities than the best CDR shuffled clone, RT3:47/4-f3. Mutant clone B6 and had an approximate Kd value of 30×10^{-9} M compared to the 69×10^{-9}M value obtained for RT3:47/4-f3.

11.3 Discussion

We have illustrated here the application of techniques that enable the isolation of large numbers of human antibodies from libraries containing 10^7–10^8 antibody V-genes and for the affinity maturation of primary isolates to obtain high affinity clones against a given target antigen. In excess of 550 different (though in many cases related) binders to the RT3 hapten were isolated and this number was limited only by the number of clones screened at each stage in the maturation scheme. Where antibodies are required that bind to specific epitopes, for example to neutralise virus activity, to activate a cell surface receptor or to stabilise the transition state of an ester hydrolysis reaction, the availability of many candidate clones increases the chance of finding one with the desired properties. Clearly the phage system can deliver such candidates more effectively than classical

Table 11.1: Summary of the Affinity Maturation of Human Binders to RT3

Library	Clone (nM)	Kd
Primary	RT3:47	N.D.
Chain Shuffled		
(Panninig)	RT3:47/4	105
(Biotin Selection)	RT3:47/f5	30
CDR Shuffled	RT3:47/4-f3	69
Mutant CDR2-VH	RT3:47/4-f3/B6	33

monoclonal antibody production with the added advantage that the candidate clones are human.

From each round of the maturation scheme clones with improved affinity were identified (Table 11.1). An ELISA based method was used to rank clones on the basis of affinity and the Kd of the chain shuffled clone RT3:47/4 was measured at 105×10^{-9}M, the CDR shuffled clone RT3:47/4-f3 was measured at 69×10^{-9}M and the biotin selected CDR2 mutant B6 was measured at 30×10^{-9}M. The in vitro maturation of the response to the RT3 hapten was harder to achieve compared to that obtained for the hapten 2-phenyl-5-oxazalone where just two rounds of mutation and selection (chain shuffling and CDR shuffling) succeeded in maturing a primary isolate with a Kd of 3.2×10^{-7}M to yield a binder with a Kd of 1.1×10^{-9}M (Marks et al., 1991). This may indicate that, as with the in vivo immune system, some antigens elicit stronger responses than others.

The use of biotinylated selections was adopted at a late stage in the maturation process. To determine the potential benefit of biotinylated selections at an earlier stage, the chain shuffled library derived from the primary isolate RT3:47 was selected twice using 0.1nM biotinylated antigen. This resulted in the isolation of a binder (designated RT3:47/F5) with a Kd of 30×10^{-9}M. It may have been possible therefore to obtain much better binders to this hapten than the mutant B6 if stringent selections had been adopted at an earlier stage (Table 11.1).

Where immunisation is possible antibodies can be directly isolated with affinities comparable to those attained with secondary response hybridomas (rodent antibodies to haptens, peptides, proteins, carbohydrates etc. have been isolated this way). However, human antibodies can only be obtained from immune sources in cases where, for example, the donor is infected with a pathogen; human antibodies to HIV (Burton et al., 1991; Barbas et al., 1992b; Barbas et al., 1993), RSV (Barbas et al., 1992c), hepatitis B (Zebedee et al., 1992), HSV, human cytomegalovirus, varicella zoster and rubella (Williamson et al., 1993) have been made from material derived from immune humans. To make human antibodies to potentially toxic agents or to self antigens (e.g. cytokines etc.) non-immune material must be used.

Unchallenged repertoires mainly consist of antibody V-genes derived from IgM antibodies and are thus diverse but contain clones of only moderate affinity. The non-immunised library used to obtain binders to RT3 was constructed from IgM derived V-genes and contained 2.9×10^7 clones. A larger library (1.6×10^8 clones) constructed from IgG derived V-genes has been shown to be unsuccessful in isolating primary binders to several different antigens; libraries of IgG derived clones reflect the recent antigenic history of the donor and are possibly too specific to use as the starting material for obtaining binders to targets that have not been encountered by the donor. IgM derived repertoires are thus a compromise between affinity and diversity.

Increasing the library size should reduce this compromise and improve the chances of obtaining primary isolates with higher affinities (Perelson and Oster, 1979). The limitations in transforming E.coli which have hampered attempts to increase library size beyond 10^9 clones have recently been circumvented by a new approach that exploits the highly efficient processes of infection and in vivo recombination (Waterhouse et al., 1993). This has made possible the construction of repertoires that are 10–100 fold larger than libraries made using conventional technologies (Griffiths et al., 1994). The system involves the site specific recombination of a variable region sequence from one replicon (a plasmid) onto a second replicon (a phage). This occurs via flanking recombination sites (lox P sites) which are introduced into the two replicons with recombination being catalysed by Cri recombinase, a protein that is supplied by the superinfected phage P1. A plasmid repertoire of, for example, 10^5 VL genes can thus be "rescued" with a phage repertoire containing 10^7 VH genes creating a repertoire of phage antibodies containing potentially 10^{12} clones. Once again, this rearrangement of gene segments to generate large phage display libraries mimics the rearrangement of gene segments that occurs in the developing B cell to generate the natural antibody repertoire. Monomeric antibody fragments with affinities in the 5–50 nanomolar range have been isolated after a single selection series from a library of 6.5×10^{10} members (Griffiths et al., 1994).

Even though these large libraries can directly yield very useful clones with affinities comparable to antibodies derived from a secondary immune response it is likely that, for therapeutic use, improvements in affinity would still be required. Some of the steps to affinity mature antibody fragments that we describe here will therefore remain a vital part of human antibody isolation programs in the future.

11.4 Conclusion

The therapeutic intervention by monoclonal antibodies can be envisaged for many different disease states via several routes; by recruiting specific components of the immune system, by blocking important receptors, by neutralising toxins or viruses or by targeting therapeutic agents to diseased cells. For these applications antibodies that elicit the least possible anti immunoglobulin responses in human recipients are required. Recent advances which enable the creation and in vitro manipulation of large repertoires of antibody fragments and the selection of rare antigen specific clones have made possible for the first time the isolation of high affinity human monoclonal antibodies to almost any desired target, including those that were previously elusive using conventional technologies (self antigens etc.).

The full realisation of the therapeutic potential of monoclonal antibodies must surely now be imminent as antibody engineering enters a new phase wherein the molecular biologist has learned to adopt strategies of the immune system to create an equivalent process in vitro.

References

Barbas, C.F., Kang, A.S., Lerner, R.A. and Benkovic, S.J. (1991) Assembly of combinatorial antibody libraries on phage surfaces: The gene III site, Proc. natl. Acad. Sci. USA *88,* 7978–7982.

Barbas, C.F. III, Bain, J.D., Hoekstra, D.M. and Lerner, R.A. (1992a) Semisynthetic combinatorial antibody libraries: a chemical solution to the diversity problem, Proc Natl Acad Sci USA *89,* 4457–61.

Barbas, C.F. III, Björling, E., Chiodi, F., Dunlop, N., Cabara, D., Jones, T.M., Zebedee, S.L., Persson, M.A.A., Nara, P.L., Norrby, E. and Burton, D.R. (1992b) Recombinant human Fab fragments neutralize human type 1 immunodeficiency virus in vitro, Proc Natl Acad Sci *89,* 9339–43.

Barbas, C.F. III, Crowe, J.E., Cababa, D., Jones, T.M., Zebedee, S.L., Murphy, B.R., Chanock, R.M. and Burton, D.R. (1992c) Human monoclonal Fab fragments derived from a combinatorial library bind to respiratory syncytial virus F glycoprotein and neutralise infectivity, Proc. Natl. Acad. Sci. USA *89,* 10164–10168.

Barbas, C.F. III, Collet, T.A., Amberg, W., Roben, P., Binley, J. M., Hoekstra, D., Cababa, D., Jones, T.M., Williamson, R.A., Pilkington, G.R., Haigwood, N.L., Cabezas, E., Satterthwaite, A. C., Sanz, I. and Burton, D.R. (1993) Molecular profile of an antibody response to HIV-1 as probed by combinatorial libraries, J Mol Biol *230,* 812–23.

Berek, C. and Milstein, C. (1988) The dynamic nature of the antibody repertoire, Immunol. Rev. *105,* 5–26.

Better, M., Chang, C.P., Robinson, R.R. and Horwitz, A.H. (1988) Escherichia coli secretion of an active chimeric antibody fragment, Science *240,* 1041–3.

Burton, D.R., Barbas, C.F. III, Persson, M.A., Koenig, S., Chanock, R.M. and Lerner, R.A. (1991) A large array of human monoclonal antibodies to type 1 human immunodeficiency virus from combinatorial libraries of asymptomatic seropositive individuals, Proc Natl Acad Sci USA *88,* 10134–7.

Chang, C.N., Landolfi, N.F. and Queen, C. (1991) Expression of antibody Fab domains on bacteriophage surfaces, J. Immunology *147,* 3610–3614.

Clackson, T., Hoogenboom, H.R., Griffiths, A.D. and Winter, G. (1991) Making antibody fragments using phage display libraries, Nature *352,* 624–628.

Foote, J. and Milstein, C. (1991) Kinetic maturation of an immune response, Nature *352,* 530–532.

Friguet, B., Chaffotte, A.F., Djavadi-Ohaniance, L. and Goldberg, M.E. (1985) Measurements of the true affinity constant in solution of antigen-antibody complexes by enzyme-linked immunosorbent assay, J Immunol Methods *77,* 305–19.

Garrard, L.J., Yang, M., O'Connell, M.P., Kelley, R.F. and Henner, D.J. (1991) Fab assembly and enrichment in a monovalent phage display system, Bio/technology *9,* 1373–1377.

Griffiths, A.D., Malmqvist, M., Marks, J.D., Bye, J.M., Embleton, M.J., McCafferty, J., Baier, M., Holliger, K.P., Gorick, B.D., Hughes-Jones, N.C., Hoogenboom, H.R. and Winter, G. (1993) Human anti-self antibodies with high specificity from phage display libraries, EMBO J *12,* 725–734.

Griffiths, A.D., Williams, S.C., Hartley, O., Tomlinson, I.M., Waterhouse, P., Crosby, W.L., Kontermann, R., Jones, P.T., Low, N., Allison, T.J., Prospero, T., Hoogenboom, H.R., Nissim, A., Cox, J.P.L., Harrison, J.L., Zaccolo, M., Gherardi, E. and Winter, G. (1994) Isolation of high affinity human antibodies directly from large synthetic repertoires, EMBO J *13,* 3245–3260.

Gussow, D. and Clackson, T. (1989) Direct clone characterization from plaques and colonies by the polymerase chain reaction, Nucleic Acids Res *17,* 4000.

Hawkins, R.E., Russell, S.J. and Winter, G. (1992) Selection of phage antibodies by binding affinity: mimicking affinity maturation, J. Mol. Biol. *226,* 889–896.

Hoogenboom, H.R., Griffiths, A.D., Johnson, K.S., Chiswell, D. J., Hudson, P. and Winter, G. (1991) Multi-subunit proteins on the surface of filamentous phage: methodologies for displaying antibody (Fab) heavy and light chains, Nucleic Acids Res *19,* 4133–4137.

Hoogenboom, H.R. and Winter, G. (1992) By-passing immunisation: Human antibodies from synthetic repertoires of germline VH gene segments rearranged in vitro, J. Mol. Biol. *227,* 381–388.

Huse, W.D. (1991) Combinatorial antibody expression libraries in filamentous phage, In: Antibody engineering. A practical approach (Borrenaeck; ed.) pp. 103–120, W H Freeman & Co, New York.

Huse, W.D., Sastry, L., Iverson, S.A., Kang, A.S., Alting, M.M., Burton, D.R., Benkovic, S.J. and Lerner, R.A. (1989) Generation of a large combinatorial library of the immunoglobulin repertoire in phage lambda, Science *246,* 1275–81.

Huse, W.M., Stinchcombe, T.J., Glaser, S.M., Starr, L., MacClean, M., Hellström, K.E. and Yelton, D.E. (1992) Application of a filamentous phage pVIII fusion protein system suitable for effcient production, screening, and mutagenesis of F(ab) antibody fragments, J of Immunol. *149,* 3914–20.

Huston, J.S., Levinson, D., Mudgett, H.M., Tai, M.S., Novotny, J., Margolies, M.N., Ridge, R.J., Bruccoleri, R.E., Haber, E., Crea, R. and Opperman, H. (1988), Protein engineering of antibody binding sites: recovery of specific activity in an anti-digoxin single-chain Fv analogue produced in Escherichia coli, Proc Natl Acad Sci USA *85,* 5879–83.

Jackson, R.H., McCafferty, J., Johnson, K.S., Pope, A.R., Roberts, A.J., Chiswell, D.J., Clackson, T.P., Griffiths, A.D., Hoogenboom, H.R. and Winter, G. (1992) Selection of variants of antibodies and other protein molecules using display on the surface of bacteriophage fd, In: Protein Engineering (Rees, Sternberg, Wetzel; eds.), pp. 277–301, Oxford University Press, Oxford.

James, K. and Bell, G.T. (1987) Human monoclonal antibody production: current status and future prospects, J. Immunol. Methods. *100,* 5–40.

Jones, P.T., Dear, P.H., Foote, J., Neuberger, M.S. and Winter, G. (1986) Replacing the complementarity-determining regions in a human antibody with those from a mouse, Nature *321,* 522–5.

Kang, A.S., Barbas, C.F. III, Janda, K.D., Benkovic, S.J. and Lerner, R.A. (1991) Linkage of recognition and replication functions by assembling combinatorial antibody Fab libraries along phage surfaces, Proc Natl Acad Sci USA *88*, 4363–6.

Kang, A.S., Jones, T.M. and Burton, D.R. (1991) Antibody redesign by chain shuffling from random combinatorial immunoglobulin libraries, Proc. Natl. Acad. Sci. USA *88*, 11120–11123.

Klotz, I.M. (1953) In: The Proteins (Neurath and Baily; eds.), p. *727*, Academic Press, New York.

Kohler, G. and Milstein, C. (1975) Continuous cultures of fused cells secreting antibody of predefined specificity, Nature *256*, 495–7.

Lowman, H.B., Bass, S.H., Simpson, N. and Wells, J.A. (1991) Selecting high-affinity binding proteins by monovalent phage display, Biochemistry *30*, 10832–10838.

Marks, J.D., Griffiths, A.D., Malmqvist, M., Clackson, T., Bye, J.M. and Winter, G. (1992) By-passing immunization: building high affinity human antibodies by chain shuffling, Bio/Technology 10, 779–783.

Marks, J.D., Hoogenboom, H.R., Bonnert, T.P., McCafferty, J., Griffiths, A.D. and Winter, G. (1991) By-passing immunization: Human antibodies from V-gene libraries displayed on phage, J. Mol. Biol. *222*, 581–597.

Marks, J.D., Ouwehand, W.H., Bye, J.M., Finnern, R., Gorick, B. D., Voak, D., Thorpe, S., Hughes-Jones, N.C. and Winter, G. (1993) Human antibody fragments specific for human blood group antigens from a phage display library, Bio/Technology *11*, 1145–1149.

McCafferty, J., FitzGerald, K.J., Earnshaw, J., Chiswell, D.J., Link, J., Smith, R. and Kenten, J. (1994) Selection and rapid purification of murine antibody fragments that bind a transition -state analog by phage display, Appl. Biochem. Biotech. *47*, 157–173.

McCafferty, J., Griffiths, A.D., Winter, G. and Chiswell, D.J. (1990) Phage antibodies: filamentous phage displaying antibody variable domains, Nature *348*, 552–4.

Munro, S.and Pelham, H.R.B. (1986) An hsp70-like protein in the ER: identity with the 78kd glucose-regulated protein and immunoglobulin heavy chain binding protein, Cell *46*, 291–300.

Neuberger, M.S., Williams, G.T., Mitchell, E.B., Jouhal, S.S., Flanagan, J.G. and Rabbitts, T.H. (1985) A hapten-specific chimaeric IgE antibody with human physiological effector function, Nature *314*, 268–70.

Orlandi, R., Gussow, D.H., Jones, P.T. and Winter, G. (1989) Cloning immunoglobulin variable domains for expression by the polymerase chain reaction, Proc Natl Acad Sci USA *86*, 3833–7.

Perelson, A.S. and Oster, G.F. (1979) Theoretical studies of clonal selection: Minimial antibody repertoire size and reliability of self non-self discrimination, J. Theor. Biol. *81*, 645–670.

Queen, C., Schneider, W.P., Selick, H.E., Payne, P.W., Landolfi, N.F., Duncan, J.F., Avdalovic, N.M., Levitt, M., Junghans, R.P. and Waldmann, T.A. (1989) A humanized antibody that binds to the interleukin 2 receptor, Proc Natl Acad Sci USA *86*, 10029–33.

Rasched, I. and Oberer, E. (1986) Ff Coliphages: Structural and Functional Relationships, Microbiol. Rev. *50*, 401–427.

Riechmann, L., Clark, M., Waldmann, H. and Winter, G. (1988) Reshaping human antibodies for therapy, Nature *332*, 323–7.

Saiki, R.K., Scharf, S., Faloona, F., Mullis, K.B., Horn, G.T., Erlich, H.A. and Arnheim, N. (1985) Enzymatic amplification of beta-globin genomic sequences and restriction site analysis for diagnosis of sickle cell anemia, Science *230*, 1350–4.

Sanger, F., Nicklen, S. and Coulson, A.R. (1977) DNA sequencing with chain-terminating inhibitors, Proc. Natl. Acad. Sci. USA *74*, 5463–7.

Sastry, L., Alting, M.M., Huse, W.D., Short, J.M., Sorge, J.A., Hay, B.N., Janda, K.D., Benkovic, S.J., Lerner, R.A. (1989) Cloning of the immunological repertoire in Escherichia coli for generation of monoclonal catalytic antibodies: construction of a heavy chain variable region-specific cDNA library, Proc Natl Acad Sci USA *86*, 5728–32.

Skerra, A. and Pluckthun, A. (1988) Assembly of a functional immunoglobulin Fv fragment in Escherichia coli, Science 240, 1038–41.

Smith, G.P. (1985) Filamentous fusion phage: novel expression vectors that display cloned antigens on the virion surface, Science *228*, 1315–1317.

Tonegawa, S. (1983) Somatic generation of antibody diversity, Nature *302*, 575–81.

Verhoeyen, M. and Riechmann, L. (1988) Engineering of antibodies, Bioessays *8*, 74–8.

Vieira, J. and Messing, J. (1987) Production of single-stranded plasmid DNA, Methods Enzymol. *153*, 3–11.

Ward, E.S., Gussow, D., Griffiths, A.D., Jones, P.T. and Winter, G. (1989) Binding activities of a repertoire of single immunoglobulin variable domains secreted from Escherichia coli, Nature *341*, 544–6.

Waterhouse, P., Griffiths, A.D., Johnson, K.S. and Winter, G. (1993) Combinatorial infection and invivo recombination: a strategy for making large phage antibody repertoires, Nucleic Acids Research *21*, 2265–66.

Williamson, R.A., Burioni, R., Sanna, P.P. and Partridge, L.J. (1993) Human monoclonal antibodies against a plethora of viral pathogens from single combinatorial libraries, Proc Natl Acad Sci USA *90*, 4141–5.

Winter, G. and Milstein, C. (1991) Man-made antibodies, Nature *349*, 293–9.

Yanisch, P.C., Vieira, J. and Messing, J. (1985) Improved M13 phage cloning vectors and host strains: nucleotide sequences of the M13mp18 and pUC19 vectors, Gene *33*, 103–19.

Zebedee, S.L., Barbas, C.F. III, Hom, Y., Caothien, R.H., Graff, R., Degraw, J., Pyati, J., LaPolla, R., Burton, D.R., Lerner, R.A. and Thronton, G.B. (1992) Human combinatorial antibody libraries to hepatitis B surface antigen, Proc. Natl. Acad. Sci. USA *89*, 3175–3179.

12 Altering the Function of Enzymes and Macromolecular Inhibitors by Phage Display

Qing Yang, Cheng-I Wang, and Charles S. Craik

12.1 Introduction

Protein engineering involves the creation of novel proteins. One approach for making new proteins involves redesigning preexisting ones for altered functions. Knowledge of structure function relationships of the protein of interest serves as a powerful starting point for engineering that protein for a particular purpose. Although any amino acid can play a significant role in maintaining the final stable structure, only a relatively small set of amino acid residues is directly responsible for function. The identification of these critical determinants of activity and the detailed study of their structural and functional contributions provide an essential database for the rational modification of a protein.

Biophysical analysis can provide critical information regarding potential roles of particular side chains in a given protein. In those cases where a significant amount of information already exists for the target protein, a directed mutagenesis approach involving the substitution of a limited number of functional residues can be applied to alter protein activity. In the absence of sufficient preestablished information regarding structure-activity relationships in a protein, one must confront the challenge of locating key residues that are important for altering function. One approach is to adapt a random mutagenesis strategy to generate large and diverse molecular repertoires. These repertoires can be searched for individual proteins with desired properties. Selected variants can then be analyzed to elucidate the structural basis for the particular function. An attractive method to aid screening of the diverse macromolecular repertoires for specific functional properties is phage display.

Phage display that genetically fuses the macromolecules of interest to the surface of bacteriophage is the most widely used procedure to date for screening repertoires of protein macromolecules. Phage display has been used to analyze peptides [Cwirla et al., 1990; Scott and Smith, 1990; Blond Elguindi, et al., 1993], antibodies [McCafferty et al., 1990; Barbas et al., 1991; Clackson et al., 1991], growth hormone variants [Bass et al., 1990; Lowman et al., 1991], DNA binding proteins [Jamieson et al., 1994; Rebar and Pabo, 1994], enzymes and protease inhibitors [Roberts et al., 1992; Roberts, et al., 1992; Pannekoek et al., 1993; Wang et al., 1994]. The peptide and antibody phage display systems have been extensively reviewed [Smith and Scott, 1993; Clackson and Wells, 1994; Gallop et al., 1994; Gordon et al., 1994; Winter et al., 1994]. The scope of this chapter will focus on various methods for altering enzymes and macromolecular inhibitors with a particular emphasis on phage display.

12.2 Background

An ultimate goal of protein engineering efforts is the *de novo* design of proteins with prescribed functions. However, until the fundamental rules for protein design become established, the *redesign* of preexisting proteins is commonly used to obtain proteins with altered functions. The field of protein engineering has witnessed continuous progress regarding protein modification and analysis techniques, which has provided greater insight on the structure-function relationship of these macromolecules. The generation of diverse libraries of protein variants and the isolation of interesting and potentially useful proteins from these libraries are two major themes for the development of new methodologies.

Protein chemistry provides a host of reagents that can be used to chemically modify one specific residue or a class of residues depending on the reactivities of the functional groups (for background, see [Lundblad, 1991; Lundblad, 1995]). However, certain reagents such as affinity labels [Shaw et al., 1965] are frequently too specific for general use while less specific reagents can only be applied to highly reactive side chains and solvent accessible sites. Furthermore, the extent of chemical modification is seldom complete leaving a heterogeneous population of modified protein. Finally, although the chemical reagents can aid in the characterization of a protein, they are rarely useful for altering the function of the target protein.

In the last decade, the development of recombinant DNA technology has offered the protein chemist new, powerful methods for modifying proteins. Site-directed mutagenesis makes it feasible to precisely and completely substitute any amino acid with any other amino acid. Heterologous expression of the mutant DNA sequence permits the production of reagent quantities of the variant protein. Coupled with X-ray crystallography or NMR spectroscopy, this new approach allows the effect of substituting a specific, single amino acid residue to be addressed in atomic detail. Many early mutagenesis studies were single amino acid substitution experiments at the active site of well-studied model enzymes (For reviews, see [Knowles, 1987; Shaw, 1987; Perona and Craik, 1995]). Clearly defined questions regarding the role of specific residues were addressed with the assumption that the contributions of these catalytically important amino acids were fairly predominant and independent. Though extremely powerful, the iterative process of mutagenize analyze-mutagenize can become time consuming and labor intensive if more than a limited number of mutants are analyzed simultaneously. Furthermore, the simplicity and the power of the mutagenesis technology rapidly outstrip the theoretical basis for structure-function relationships for a particular protein. Without extensive *a priori* knowledge of the protein, it is seldom clear what amino acids are appropriate targets for mutagenesis.

The sophistication of contemporary protein modification techniques has provided unprecedented opportunities to create altered or even completely novel functions of proteins, especially enzymes. Although enzymes are highly efficient and specialized, it is sometimes possible to alter or actually improve them for specific reactions. An example of this type of redesign can be seen in naturally occurring enzymes with similar sequences and three-dimensional structures that possess distinct functions. Eventually it should be possible to create synthetic proteins with novel catalytic activities from preexisting structures with sufficient mutations. However, it is not obvious how to do this in most cases. Although there are examples where a few amino acid substitutions have resulted in drastic changes in enzyme function while maintaining sufficient catalytic activity [Higaki et al., 1990; Sloane et al., 1991; Wilks et al., 1988; Wells et al., 1987], it is often found that many substitutions are required to successfully alter the properties of an enzyme [Wells and Estell, 1988; Hedstrom et al., 1992]. Therefore, the protein engineer faces the formidable task of locating and identifying the amino acids necessary for altering the function of a protein.

This challenge has stimulated two innovative advances in the field:

(1) The generation of molecular repertoires with sufficient diversity for altered function to be represented;

(2) The development of screens or selections to effectively sort through the large repertoires.

It is now common practice to generate large diverse DNA libraries that express variant proteins due to simultaneous and complementary advances in: 1) automated solid phase oligonucleotide synthesis; 2) DNA cloning; and 3) heterologous expression of foreign DNA in bacteria. Figure 12.1 gives an overview of various mutagenesis strategies for altering a protein. In one method referred to as "cassette mutagenesis" [Botstein and Shortle, 1985; Wells et al., 1985], one or a few codons in the gene encoding the target protein are replaced with nucleotide mixtures encoding all twenty amino acids. This creates a library of variant proteins with random mutations in defined positions. Other methods, including

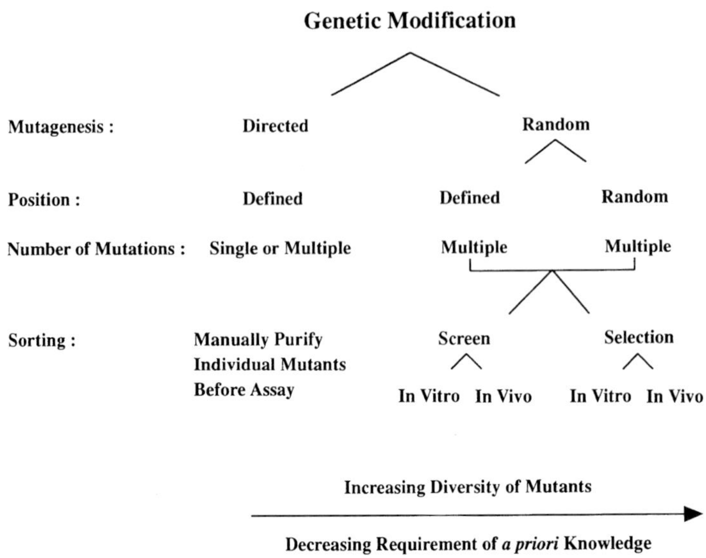

Figure 12.1: Recombinant Methods for Protein Modification

mutator strains [McCoy and Khorana, 1983], chemical mutagenesis of DNA [Kadonaga and Knowles, 1985], enzymatic misincorporation of nucleotides [Shortle and Lin, 1985], doped DNA synthesis [Hermes et al., 1989], error prone PCR [Leung et al., 1989], or DNA shuffling [Stemmer, 1994] produce mutations randomly distributed along the target gene (see Figure 12.1 for comparison). The diversity of these libraries is only limited by the transformation efficiency of *E.coli*, usually on the order of 10^8 to 10^9 colonies per microgram of DNA. The absence of *a priori* knowledge for altering the function of a protein can be overcome by increasing the size and diversity of the variant library and efficiently searching the library for the variant that exhibits the desired activity.

The approaches that have been established for sorting through combinatorial molecular repertoires can be grouped into two broad categories: selections and screens. A screen will provide relative scores for a spectrum of mutants under particular assay conditions. It is then up to the researcher to decide which mutant is worthy of further study. A selection procedure, on the other hand, will enrich a subset of mutants that meet the criteria of the experimental design while the rest of the population is lost or discarded. Depending upon the sensitivity and the dynamic range of the selection, the activity of the variant protein determines which mutants survive and undergo further study. Screens and selections can also be defined based on the location of the interaction, either *in vivo* or *in vitro*. Representative examples of each system are listed in Table 12.1.

Table 12.I: Methods for Analyzing Macromolecule Repertoires

		Site of Interaction	
		in Vivo	In Vitro
Method	Screen	• Yeast two -hybrid system [Chien et al., 1991]	• DNA cloning by primer hybridization or antibody recognition [Sambrook et al. eds., 1989] • Monoclonal antibody [Kohler and Milstein, 1975] • Peptides-on-Pins [Geysen et al., 1984] • "Tea-bag" [Houghten, 1985]
	Selection	• cDNA cloning by complementation [Chang et al., 1978] • Activity selection [Evnin et al., 1990]	• SELEX [Tuerk and Gold, 1990] • Aptamer [Bock et al., 1992] • Peptides-on-Plasmids [Cull et al., 1992] • Phage display [Smith, 1985]

A mutant library saturated at position 71 of β-lactamase has been characterized by antibiotic resistance in *E.coli*. The selection revealed that this position was important for protein stability but was not essential for catalysis [Schultz et al., 1986]. Recently, *in vitro* homologous recombination (DNA shuffling) was used to generate a library of β-lactamase mutants from which a mutant was selected with 32,000-fold increase of minimum inhibitory concentration against the antibiotic cefotaxime [Stemmer, 1994]. The selection of functional variants of lambda repressor from a library of dimer interface mutants showed that the number and type of substitutions allowed at each position were extremely variable [Reidhaar and Sauer, 1988]. Knowles and coworkers have made library of mutant isolated that incorporated mutations that could improve the catalytic efficiency of a sluggish mutant [Hermes et al., 1990] . An E. coli auxotroph has been used to select trypsin variants that prefer arginine or lysine substrates from a binding pocket library [Evnin et al., 1990; Perona et al., 1993]. A partially randomized library involving eleven codons of thymidine kinase has been subjected to genetic selection. Conserved amino acids that may be important for thymidine kinase catalysis were identified [Munir et al., 1992; Munir et al., 1993]. Saturated mutagenesis has also been applied to conserved residues in thymidylate synthase [Climie et al., 1990]. Catalytically active mutants were isolated by complementation using a Thy- strain of *E.coli*. The results point to a high degree of flexibility of thymidylate synthase in accommodating function with structural change. Recently, genetic selections have also been used to characterize libraries of potential catalytic antibodies [Lesley et al., 1993; Smiley and Benkovic, 1994]

These encouraging examples illustrate the great potential of using *in vivo* selections to sort through libraries of variant proteins. However, the particular activity under selective pressure has to be related to a vital metabolic step in the host organism. This restriction greatly reduces the number of enzymes or other proteins that may be the target of *in vivo* selection. Besides, our limited understanding of the overall metabolism of the host organism also overshadows the efforts to design and construct such experimental systems. Therefore, the development of *in vitro* selection systems provides a promising alternative to overcome the inherent difficulties in developing a practical genetic selection.

The *in vitro* selection systems for sorting through large nucleic acid libraries [Szostak, 1992; Bartel and Szostak, 1993; Lorsch and Szostak, 1994)] take advantage of the polymerase chain reaction (PCR) to directly amplify selected individual variants *in vitro*. Since no equivalent of the polymerase chain reaction for amplifying nucleic acids exists for directly amplifying proteins, methods that link the function (phenotype) and identity (genotype) of the protein of interest are required for protein repertoires.

Techniques have been developed that couple chemical tags to peptide sequences [Brenner and Lerner, 1992][Needels et al., 1993]. This facilitates rapid identification of a particular member in a library. However, peptides alone cannot be amplified *in vivo*. Therefore, a more versatile and efficient system for screening protein libraries needs to incorporate both the identification and amplification of individual members. One way to achieve this is to link the macromolecule of interest to a biological carrier, which can be uniquely identified and reproduced.

One of these methods is called "peptides-on-plasmids"[Cull et al., 1992; Schatz, 1993]. A LacI-peptide fusion protein is specifically bound to the Lac operator sequence in the plasmid that encodes the peptide sequence. The protein-DNA complex can be isolated and passed through an immobilized ligand matrix to retain those peptides that bind the ligand. After elution from the matrix, the plasmid DNA can be amplified and sequenced. This system allows intracellular expression of peptides on the C terminus of the *Lac repressor*. The substrate specificity of an E.coli biotinylation enzyme has been mapped with peptide libraries based on this system [Schatz, 1993]. Other systems use the surface protein on bacteria or virus as the expression platform [Little et al., 1993]. The intact bacteria or virus particle becomes the vehicle to display foreign proteins or peptides of interest. *E.coli* surface protein LamB [Charbit et al., 1988] and OmpA [Francisco et al., 1992] have been tested for this purpose. Although each of these methods offer unique insights into the particular protein of interest, the approach that has gained wide acceptance for sorting through libraries of peptide or protein variants is filamentous phage display.

12.3 Filamentous Phage Display System

The structure and life cycle of filamentous phage M13 have been the subject of extensive investigation as model systems for virology. The detailed understanding of M13 filamentous phage and its widespread use in recombinant DNA technology permitted the rapid development of it as a display vehicle for foreign proteins and peptides [Smith, 1988]. M13 phage can be rapidly propagated through the infection of *E.coli*, resulting in massive amplification of phage clones. The phage particles are stable in a wide range of pH and salt concentrations, allowing flexible conditions for washing and elution procedures. The phage particle displaying the protein of interest can bind to an immobilized ligand matrix through the foreign protein domain and be specifically enriched. The process of phage binding, washing and elution on a solid support is given the name "panning" [Parmley and Smith, 1988]. Multiple panning cycle is a definite advantage for the phage display system. Even if the enrichment of a single round of panning is not very significant, strong binders may still be identified through repetitive rounds of panning. This productive loop greatly increases the sensitivity of the in vitro selection and enables rare individual clones with target binding properties to be isolated from large, random libraries with up to 10^{8-9} individual clones.

In the phage display system, the selective force is the binding interaction between the phage and ligand in solution. This artificial selection pressure allows one to manipulate various conditions to enrich the desired phage populations. Parameters like pH, ionic strength, temperature, duration of binding incubation, number of wash steps and concentration of competing ligands can affect the stringency of the panning experiment. For example, variation of the incubation period with competing ligands can be used to select for phage with slow K_{off} rates [Hawkins et al., 1992]. The flexibility of the phage display system towards selection conditions makes it easier to isolate different variant populations from a single pool under adjustable stringencies. Compared with the "life or death" outcome of the genetic selection, this approach can often provide more quantitative information on the nature of the binding interaction. Another distinct advantage of the phage display system is that it poses few constraints on the physical and chemical nature of the ligand and matrix. In a genetic selection, the desired scheme has to fit within the metabolic constraints of the host organism. This limitation greatly reduces the number of possible ligands with unnatural chemical moieties.

The ease to generate and reamplify large and diverse pools of fusion phage variants, the versatility towards binding conditions and ligands, plus the unique physical linkage between recognition and replication are the foundations of the phage display technique. Several experimental systems have been established to display enzymes and macromolecular inhibitors on phage to try to explore the potential of this technology. The enzymes that have been displayed include alkaline phosphatase [McCafferty et al., 1991], trypsin [Corey et al., 1993], β-lactamase [Soumillion et al., 1994] and prostate specific antigen (PSA) [Eerola et al., 1994]. The macromolecular protease inhibitors being displayed are BPTI [Roberts et al., 1992], PAI-1 [Pannekoek et al., 1993], APPI [Clackson and Wells, 1994] and ecotin [Wang et al., 1994].

12.4 Enzymes Displayed on Phage

12.4.1 Trypsin

Trypsin is a member of the serine protease family. Whenever a new and powerful technique emerges, proteases have been among the first proteins to be tested with the novel approach. This, in turn, has resulted in serine proteases being one of the best studied classes of enzymes regarding substrate specificity and catalytic mechanism. The catalytic residues and primary determinants of substrate specificity have been mutated individually to address the role of the various amino

Figure 12.2: Phage Display of Trypsin

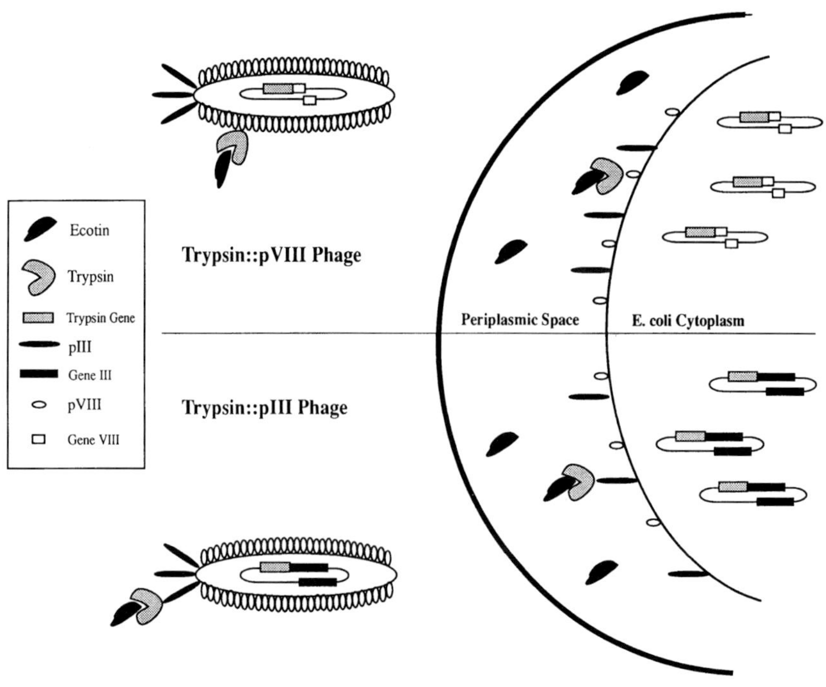

acids [Craik et al., 1985; Craik et al., 1987; Corey and Craik, 1992]. A binding pocket library has also been made and subjected to *in vivo* genetic selection in auxotrophic strains of *E.coli* [Evnin et al., 1990; Perona et al., 1993]. To further investigate the determinants of its substrate specificity and introduce novel specificities, trypsin was displayed on the surface of filamentous phage [Corey et al., 1993].

The gene encoding mature rat anionic trypsin was fused via a di-glycine linker to the C-terminal domain of geneIII or the entire geneVIII of bacteriophage M13 coat protein. The expression of the trypsin coat protein fusion is regulated by a Tac promoter and targeted for secretion into the periplasm of *E.coli* by a HisJ signal peptide. Removal of the signal peptides by leader peptidase results in producing active trypsin since the zymogen peptide is not present. The fusion protein is assembled onto the surface of the filamentous phage. The trypsin on phage forms a complex with ecotin, a protease inhibitor in the periplasm of *E. coli* [Chung et al., 1983; McGrath et al., 1991] (Figure 12.2). Active trypsin phage are purified from ecotin after extensive dialysis at pH2.2 against a 300 kD molecular weight cut-off membrane.

The trypsin phage can be stained with anti-trypsin antibody after electrophoresis and immuno-blotting, indicating that trypsin is expressed on phage. The trypsin phage is active against small synthetic substrates such as benzyloxycarbonyl-Gly-Pro-Arg-aminomethyl-coumarin (Z GPR-AMC) and is inhibited by the tight binding trypsin inhibitor bovine pancreatic trypsin inhibitor (BPTI). These results show that trypsin on phage is successfully secreted to the periplasm, can be correctly folded and achieve normal catalytic activity while fused to geneIII or geneVIII proteins (pIII or pVIII). The fusion phage possess similar steady-state kinetic parameters (kcat and K_m) to that of wild type trypsin. Trypsin phage can be selectively enriched from background M13 phage by ecotin immobilized on petri dishes or acrylic beads by at least one thousand fold in a single round of panning. The binding interaction is independent of the catalytic activity since the inactive variant, trypsin D102N displayed on phage can also be captured by the same ligand. The catalytic efficiency of this variant is four orders of magnitude lower than that of trypsin [Corey and Craik, 1992].

Studying trypsin in such a system may provide new insight into the interplay of structure and specificity of trypsin. A goal of these efforts is to generate trypsin mutants with desired novel specificities and to eventually make more intelligent predictions in the *de novo* design of a macromolecule.

12.4.2 Alkaline Phosphatase

Alkaline phosphatase (AP) is a dimeric enzyme in the periplasm of *E.coli*. McCafferty et al. have expressed functional AP and an R166A mutant on the surface of bacteriophage by inserting the *E.coli* AP gene *(phoA)* into the vector fd-tet-DOG1 after the geneIII signal sequence [McCafferty et al., 1991]. The construction generates a mature fusion consisting of five amino acids derived from geneIII at the N terminus, followed by a complete AP monomer, fused via a tri-alanine linker to the mature geneIII protein.

The AP-pIII fusion can be detected by anti-pIII and anti-AP antibodies. Size separation by ultrafiltration confirms that the fusion protein is bound to the phage. The catalytic activity has been demonstrated and the kinetic parameters have been measured for both wild type and a mutant form of the phage-enzyme fusion. The K_m value for the AP-phage is about 9 fold higher than that of the soluble enzyme. Replacement of an arginine by an alanine at the active site (R166A) has a qualitatively similar effect on the kinetic properties of both soluble and phage-bound enzymes. Phage particles expressing both mutant and wild type AP can be retained on arsenate-Sepharose and can be eluted with inorganic phosphate. This result demonstrates the feasibility of using a competitive inhibitor to capture phage-bound enzymes.

12.4.3 β-Lactamase

One of the intrinsic limitations of the phage display method for selecting enzyme variants is its sole dependence on binding. However, a tight enzyme substrate binding interaction does not necessarily transform into high catalytic efficiency. To overcome this problem, Soumillion et. al. have developed a system to display β-lactamase on phage and used a irreversible, mechanism based inhibitor as the ligand to select for active phage-bound enzyme against an active site mutant (S70A) [Soumillion et al., 1994]. The inhibitor specifically reacts with the active site of *b*-lactamase, forming a covalent bond with Ser70 of the enzyme.

The target enzyme is RTEM β-lactamase, a secreted monomeric enzyme. The β-lactamase is fused to the N terminus of pIII via a seven residue spacer, which includes a factor Xa cleavage site. The β-lactamase is expressed on phage with full activity. The mechanism-based inhibitor is incorporated into one end of a bifunctional ligand with a disulfide linked biotin moiety at the other end. After incubation with this bifunctional ligand, the phage-enzyme-inhibitor complex is captured by streptavidin-coated magnetic beads. The phage encoding active β-lactamase are subsequently released by factor Xa cleavage or by DTT reduction. The phage displaying wild type β-lactamase were enriched over the phage displaying the S70A variant. The factor Xa cleavage resulted in a 50 fold enrichment while the DTT reduction resulted in only 8 fold enrichment. The low level of enrichment may result from inefficient release of covalently bound phage or non-discriminatory binding of phage displaying active β-lactamase. The incorporation of a selection based on activity is a significant step that brings catalysis into the *in vitro* binding selection and should be applicable to other enzyme systems. It may eventually facilitate the isolation of active enzyme variants with altered specificities or other novel properties.

12.5 Macromolecular Protease Inhibitors Displayed on Phage

12.5.1 BPTI

Bovine pancreatic trypsin inhibitor (BPTI) is the first macromolecular inhibitor displayed on M13 phage. Roberts et al. have created a library of BPTI variants on phage to select for specific inhibitors of human neutrophil elastase (HNE) [Roberts et al., 1992; Roberts et al., 1992; Markland et al., 1991].

The gene for mature BPTI is inserted between the signal peptide and mature protein products of geneIII. The library of BPTI phage consists of up to 1,000 different mutants with hydrophobic residues at position 15–19 of the reactive loop in the inhibitor reactive site. The library is panned with HNE-agarose beads and fractionated sequentially with buffers of decreasing pH. After three rounds of panning, some mutants are prevalent in the population. All of them show significantly higher affinity for HNE. One of them, the most potent anti-HNE, Kunitz type inhibitor described to date, has affinity over six orders of magnitude higher than that of the parental protein. The power of phage display for iterative affinity selection is clearly shown here.

12.5.2 PAI-1

Pannekoek and co-workers have expressed the protease inhibitor, human plasminogen-activator inhibitor 1 (PAI-1) on phage [Pannekoek et al., 1993]. The fusion construct includes the *pelB* leader peptide, an extra N-terminal sequence AQVKL plus the mature PAI-1, a five residue spacer GGGGS and C-terminal half of pIII. The phage-displayed PAI-1 specifically binds to anti-PAI-1 antibody and retains its capacity to form complexes with its target, tissue-type plasminogen activator (t-PA). It also inhibits the aminolytic activity of t-PA in a dose-dependent fashion. A phage library containing predominantly single point mutations throughout the entire gene was constructed by error-prone PCR. This provided an alternative approach for introducing mutations in a target protein for subsequent phage display. Because PAI-1 is a key regulatory protein in the fibrinolytic system, the opportunity exists for isolating a variant PAI-1 that can distinguish among the various binding interactions of the pathway.

12.5.3 Ecotin

Ecotin is a dimeric serine protease inhibitor found in the periplasm of *E.coli*, where each unit of the dimer contains 142 amino acids [Chung et al., 1983; McGrath et al., 1991]. Ecotin has been found to inhibit pancreatic serine proteases of a broad range of specificity but not any known proteases from *E.coli* [Chung et al., 1983]. Recently, ecotin has also been found to be a highly potent anticoagulant and a reversible tight-binding inhibitor of human factor Xa [Seymour et al., 1994]. Ecotin belongs to the "substrate-like" class of inhibitors with Met 84 at the reactive site (the P1 site) [McGrath et al., 1991; Laskowski and Kato, 1980]. A crystal structure of ecotin complexed with trypsin showed that two trypsin molecules bind to an ecotin dimer in a 2-fold symmetry [McGrath et al., 1994]. In addition to the interactions through a primary site that includes the reactive site loop, ecotin makes a total of 9 hydrogen bonds to trypsin through a secondary binding site located at the distal end of ecotin relative to the reactive site. Modeling studies with ecotin and other proteases including chymotrypsin and elastase indicates that similar interactions could occur, along with other unique contacts. The chelation of a target protease through the two binding sites is a unique feature of ecotin since most serine protease inhibitors interact with their target proteases predominately through their reactive site loop [Bode and Huber, 1992]. Wang and colleagues have developed a system for displaying a repertoire of ecotin mutants on the surface of bacteriophage to allow for selection of high affinity ecotin variants toward specific proteases, especially for those with therapeutic importance [Wang et al., 1994].

The ecotin gene along with its signal sequence from the genomic clone is fused to the C-terminal domain of geneIII via a tri-glycine linker. The fusion gene is expressed in a pBluescript vector under transcriptional control by the lac promoter. The fusion phage react with anti-ecotin antibodies and are capable of inhibiting the proteolytic activity of trypsin with an affinity similar to that of the free ecotin. Ecotin phage can be enriched 10^4 to 10^5 fold by immobilized trypsin from a wild type M13 phage background. A phage library where the reactive site residues 84 (P1) and 85 (P1′) of ecotin are replaced with all possible substitutions was generated by oligonucleotide-directed mutagenesis. The phage library was panned for three rounds against human u-PA immobilized on polystyrene dishes. Individual phage clones from each round of binding selection were sequenced to determine their identity. A consensus pattern clearly emerges after each round of selection. The positively charged amino acids arginine (Arg) and lysine (Lys) are preferred at both of the positions. The M84R/M85R mutant predominated in the population after the third round. Ecotin variants found in the third round of selection were expressed, purified and characterized for their inhibitory affinity against u-PA. The Ki of the ecotin mutant M84R/M85R is approximately one nanomolar against u-PA, nearly three orders of magnitude lower than that of wild type ecotin. This promising start illus-trates the potential of the phage display system for engineering novel inhibitory specificity in ecotin against other therapeutically important proteases.

12.6 Conclusions

The phage display technique is an innovative tool for "engineering" enzymes and their macromolecular inhibitors. It offers new opportunities to investigate the combinatorial interactions among the functionally important residues of a target macromolecule. It provides an efficient means to sort through diverse libraries of mutants based on the *in vitro* binding properties of the variant protein. The molecular diversity that the current phage display technique can generate and process permits *in vitro* applied molecular evolution [Kauffman, 1992] experiments, by which novel catalytic or inhibitory functions can be searched within large phage repertoires. The above examples of enzymes and protease inhibitors successfully displayed on phage suggest that this technique may be applied to other experimental systems as well. The continued development of *in vitro* selection techniques that are highly discriminating and that incorporate activity based selection criteria will have a profound impact on the field of protein engineering.

Acknowledgment

The authors are supported by National Science Foundation Grant MCB-9219806 (to C.S.C.) and NIH Pharmaceutical Chemistry, Pharmacology, Toxicology Training Grant GM07175 (to Q.Y.).

References

Barbas, C.F., III, Kang, AS., Lerner, R.A. and Benkovic, S.J. (1991) Assembly of combinatorial antibody libraries on phage surfaces: the gene III site, Proc Natl Acad Sci USA *88*, 7978–82.

Bartel, D.P. and Szostak, J.W. (1993) Isolation of new ribozymes from a large pool of random sequences [see comment], Science *261*, 1411–8.

Bass, S., Greene, R. and Wells, J.A. (1990) Hormone phage: an enrichment method for variant proteins with altered binding properties, Proteins *8*, 309–14.

Blond-Elguindi, S., Cwirla, S.E., Dower, D.J., Lipshutz, R.J., Sprang, S.R., Sambrook, J.F. and Gething, M.J. (1993) Affinity panning of a library of peptides displayed on bacteriophages reveals the binding specificity of BiP, Cell *75*, 717–28.

Bock, L.C., Griffin, L.C., Latham, J.A., Vermaas, E.H. and Toole, J.J. (1992) Selection of single-stranded DNA molecules that bind and inhibit human thrombin, Nature *355,* 564–6.

Bode, W. and Huber, R. (1992) Natural protein proteinase inhibitors and their interaction with proteinases, Eur J Biochem *204,* 433–51.

Botstein, D. and Shortle, D. (1985) Strategies and applications of *in vitro* mutagenesis, Science *229,* 1193–201.

Brenner, S. and Lerner, R.A. (1992) Encoded combinatorial chemistry, Proc Natl Acad Sci USA *89,* 5381–3.

Chang, A.C., Nunberg, J.H., Kaufman, R.J., Erlich, H.A., Schimke, R.T. and Cohen, S.N. (1978) Phenotypic expression in *E.coli* of a DNA sequence coding for mouse dihydrofolate reductase, Nature *275,* 617–24.

Charbit, A., Molla, A., Saurin, W. and Hofnung, M. (1988) Versatility of a vector for expressing foreign polypeptides at the surface of gram-negative bacteria, Gene *70,* 181–9.

Chien, C.T., Bartel, P.L., Sterglanz, R. and Fields, S. (1991) The two-hybrid system: a method to identify and clone genes for proteins that interact with a protein of interest, Proc Natl Acad Sci USA *88,* 9578–82.

Chung, C.H., Ives, H.E., Almeda, S. and Goldberg, A.L. (1983) Purification from Escherichia coli of a periplasmic protein that is a potent inhibitor of pancreatic proteases, J Biol Chem *258,* 11032–8.

Clackson, T. and Wells, J.A. (1994) *in vitro* selection from protein and peptide libraries, Trends Biotechnol. *12,* 173–184.

Clackson, T., Hoogenboom, H.R., Griffiths, A.D. and Winter, G. (1991) Making antibody fragments using phage display libraries, Nature *352,* 624–8.

Climie, S., Ruiz, P.L., Gonzales, P.D., Prapunwattana, P., Cho, S.W., Stroud, R. and Santi, D.V. (1990) Saturation site-directed mutagenesis of thymidylate synthase, J Biol Chem *265,* 18776–9.

Corey, D.R., Shiau, A.K., Yang, Q., Janowski, B.A. and Craik, C. S. (1993) Trypsin display on the surface of bacteriophage, Gene *128,* 129–34.

Corey, D.R. and Craik, C.S. (1992) An investigation into the minimum requirements for peptide hydrolysis by mutation of the catalytic triad of trypsin, J. Am. Chem. Soc. *114,* 1784–1790.

Craik, C.S., et al., (1985). Redesigning trypsin: alteration of substrate specificity. Science, *228,* 291–7.

Craik, C.S., et al., (1987). The catalytic role of the active site aspartic acid in serine proteases. Science, *237,* 909–13.

Cull, M.G., Miller, J.F. and Schatz, P.J. (1992) Screening for receptor ligands using large libraries of peptides linked to the C terminus of the lac repressor, Proc Natl Acad Sci USA *89,* 1865–9.

Cwirla, S.E., Peters, E.A., Barrett, R.W. (1990) Peptides on phage: a vast library of peptides for identifying ligands, Proc Natl Acad Sci USA *87,* 6378–82.

Eerola, R., Saviranta, P., Lilja, H., Pettersson, K., Lovgren, T. and Karp, M. (1994) Expression of prostate specific antigen on the surface of a filamentous phage, Biochem Biophys Res Commun *200,* 1346–52.

Evnin, L.B., Vasquez, J.R. and Craik, C.S. (1990) Substrate specificity of trypsin investigated by using a genetic selection, Proc Natl Acad Sci USA *87,* 6659–63.

Francisco, J.A., Earhart, C.F. and Georgiou, G. (1992) Transport and anchoring of beta-lactamase to the external surface of Escherichia coli, Proc Natl Acad Sci USA *89,* 2713–7.

Gallop, M.A., Barrett, R.W., Dower, W.J., Fodor, S.P. and Gordon, E.M. (1994) Applications of combinatorial technologies to drug discovery. 1. Background and peptide combinatorial libraries, J Med Chem *37,* 1233–51.

Geysen, H.M., Meloen, R.H. and Barteling, S.J. (1984) Use of peptide synthesis to probe viral antigens for epitopes to a resolution of a single amino acid, Proc Natl Acad Sci USA *81,* 3998–4002.

Gordon, E.M., Barrett, R.W., Dower, W.J., Fodor, S.P. and Gallop, M.A. (1994) Applications of combinatorial technologies to drug discovery. 2. Combinatorial organic synthesis, library screening strategies, and future directions, J Med Chem *37*, 1385–401.
Hawkins, R.E., Russell, S.J. and Winter, G. (1992) Selection of phage antibodies by binding affinity. Mimicking affinity maturation, J Mol Biol *226*, 889–96.
Hedstrom, L., Szilagyi, L. and Rutter, W.J. (1992) Converting trypsin to chymotrypsin: the role of surface loops, Science *255*, 1249–53.
Hermes, J.D., Blacklow, S.C. and Knowles, J.R. (1990) Searching sequence space by definably random mutagenesis: improving the catalytic potency of an enzyme, Proc Natl Acad Sci USA *87*, 696–700.
Hermes, J.D., et al., (1989) A reliable method for random mutagenesis: the generation of mutant libraries using spiked oligodeoxyribonucleotide primers, Gene *84*, 143–51.
Higaki, J.N., Haymore, B.L., Chen, S., Fletterick, R.J. and Craik, C. S.(1990) Regulation of serine protease activity by an engineered metal switch, Biochemistry *29*, 8582–6.
Houghten, R.A. (1985) General method for the rapid solid-phase synthesis of large numbers of peptides: specificity of antigen antibody interaction at the level of individual amino acids, Proc Natl Acad Sci USA *82*, 5131–5.
Jamieson, A.C., Kim, S.H. and Wells, J.A. (1994) *in vitro* selection of zinc fingers with altered DNA-binding specificity, Biochemistry *33*, 5689–95.
Kadonaga, J.T. and Knowles, J.R. (1985) A simple and efficient method for chemical mutagenesis of DNA, Nucleic Acids Res *13*, 1733–45.
Kauffman, S.A. (1992) Applied molecular evolution, J Theor Biol *157*, 1–7.
Knowles, J.R. (1987) Tinkering with enzymes: what are we learning? Science *236*, 1252–8.
Kohler, G. and Milstein, C. (1975) Continuous cultures of fused cells secreting antibody of predefined specificity, Nature *256*, 495–7.
Laskowski, M.J. and Kato, I. (1980) Protein inhibitors of proteinases, Annu Rev Biochem *49*, 593–626.
Lesley, S.A., Patten, P.A. and Schultz, P.G. (1993) A genetic approach to the generation of antibodies with enhanced catalytic activities, Proc Natl Acad Sci USA *90, 1160–5.*
Leung, D.W., Chen, E.Y. and Goeddel, D.V. (1989) A method for random mutagenesis of a defined DNA segment using a modified polymerase chain reaction, Technique *1*, 11–15.
Little, M., Fuchs, P., Breitling, F. and Dubel, S. (1993) Bacterial surface presentation of proteins and peptides – an alternative to phage technology, Trends Biotechnol. *11*, 3–5.
Lorsch, J.R. and Szostak, J.W. (1994) *in vitro* evolution of new ribozymes with polynucleotide kinase activity, Nature *371*, 31–6.
Lowman, H.B., Bass, S.H., Simpson, N. and Wells, J.A. (1991) Selecting high-affinity binding proteins by monovalent phage display, Biochemistry *30*, 10832–8.
Lundblad, R.L. (1991) Chemical reagents for protein modification, 2nd ed., CRC Press, Boca Raton, Florida.
Lundblad, R.L. (1995) Techniques in protein modification, CRC Press, Boca Raton, Florida.
Markland, W., Roberts, B.L., Saxena, M.J., Guterman, S.K. and Ladner, R.C. (1991) Design, construction and function of a multicopy display vector using fusions to the major coat protein of bacteriophage M13, Gene *109*, 13–9.
McCafferty, J., Griffiths, A.D., Winter, G. and Chiswell, D.J. (1990) Phage antibodies: filamentous phage displaying antibody variable domains, Nature *348*, 552–4.
McCafferty, J., Jackson, R.H. and Chiswell, D.J. (1991) Phage enzymes: expression and affinity chromatography of functional alkaline phosphatase on the surface of bacteriophage, Protein Eng *4*, 955–61.
McCoy, J.M. and Khorana, H.G. (1983) Introduction and characterization of amber mutations in the bacteriorhodopsin gene, J Biol Chem *258*, 8456–61.
McGrath, M.E., Erpel, T., Bystroff, C. and Fletterick, R.J. (1994) Macromolecular chelation as an improved mechanism of protease inhibition: structure of the ecotin-trypsin complex, Embo J *13*, 1502–7.

McGrath, M.E., Hines, W.M., Sakanari, J.A., Fletterick, R.J. and Craik, C.S. (1991) The sequence and reactive site of ecotin. A general inhibitor of pancreatic serine proteases from Escherichia coli, J Biol Chem *266,* 6620–5.

Munir, K.M., French, D.C. and Loeb, L.A. (1993) Thymidine kinase mutants obtained by random sequence selection, Proc Natl Acad Sci USA *90,* 4012–6.

Munir, K.M., French, D.C., Dube, D.K. and Loeb, L.A. (1992) Permissible amino acid substitutions within the putative nucleoside binding site of herpes simplex virus type 1 encoded thymidine kinase established by random sequence mutagenesis [corrected] [published erratum appears in J Biol Chem 1992 Jul 25; 267(21):15258], J Biol Chem *267,* 6584–9.

Needels, M.C., Jones, D.G., Tate, E.H., Heinkel, G.L., Kocherspenger, L.M., Dower, W.J., Barrett, R.W. and Gallop, M.A. (1993) Generation and screening of an oligonucleotide-encoded synthetic peptide library, Proc Natl Acad Sci USA *90,* 10700–4.

Pannekoek, H., van Meijer, M., Schleef, R.R., Loskutoff, D. J. and Barbas, C.F., III(1993) Functional display of human plasminogen-activator inhibitor 1 (PAI-1) on phages: novel perspectives for structure-function analysis by error-prone DNA synthesis, Gene *128,* 135–40.

Parmley, S.F. and Smith, G.P. (1988) Antibody-selectable filamentous fd phage vectors: affinity purification of target genes, Gene *73,* 305–18.

Perona, J.J. and Craik, C.S. (1995) Structural basis of substrate specificity in the serine proteases, Protein Science, *4,* 337–360.

Perona, J.J., Evnin, L.B. and Craik, C.S. (1993) A genetic selection elucidates structural determinants of arginine versus lysine specificity in trypsin, Gene *137,* 121–6.

Rebar, E.J. and Pabo, C.O. (1994) Zinc finger phage: affinity selection of fingers with new DNA-binding specificities, Science *263,* 671–3.

Reidhaar, O.J. and Sauer, R.T. (1988) Combinatorial cassette mutagenesis as a probe of the informational content of protein sequences, Science *241,* 53–7.

Roberts, B.L., et al., (1992) Directed evolution of a protein: selection of potent neutrophil elastase inhibitors displayed on M13 fusion phage, Proc Natl Acad Sci USA *89,* 2429–33.

Roberts, B.L., et al., (1992) Protease inhibitor display M13 phage: selection of high-affinity neutrophil elastase inhibitors, Gene *121,* 9–15.

Sambrook, J., Fritsch, E.F. and Maniatis, T.; eds. (1989) Molecular cloning: a laboratory manual. 2nd ed., Cold Spring Harbor Laboratory Press, Plainview, New York.

Schatz, P.J. (1993) Use of peptide libraries to map the substrate specificity of a peptide-modifying enzyme – a 13 residue consensus peptide specifies biotinylation in escherichia-coli, Bio-technol. 11 , 1138–1143.

Schultz, S.C. and Richards, J.H. (1986) Site-saturation studies of beta-lactamase: production and characterization of mutant beta-lactamases with all possible amino acid substitutions at residue 71, Proc Natl Acad Sci USA *83,* 1588–92.

Scott, J.K. and Smith, G.P. (1990) Searching for peptide ligands with an epitope library, Science *249,* 386–90.

Seymour, J.L., Lindqvist, R.N., Dennis, M.S., Moffat, B., Yansura, D., Reilly, D., Wessinger, M.E. and Lazarus, R.A. (1994) Ecotin is a potent anticoagulant and reversible tight-binding inhibitor of factor Xa, Biochemistry *33,* 3949–58.

Shaw, E., Mares-Guia, M. and Cohen, W. (1965) Evidence for an active-center histidine in trypsin through the use of a specific reagent, 1-chloro-3-tosylamido-7-amino-2-heptanone, the chloromethyl ketone derived from Na-toysl-L-lysine, Biochemistry *4,* 2219.

Shaw, W.V. (1987) Protein engineering. The design, synthesis and characterization of factitious proteins, Biochem J *246,* 1–17.

Shortle, D. and Lin, B. (1985) Genetic analysis of staphylococcal nuclease: identification of three intragenic "global" suppressors of nuclease-minus mutations, Genetics *110,* 539–55.

Sloane, D.L., Leung, R., Craik, C.S. and Sigal, E. (1991) A primary determinant for lipoxygenase positional specificity, Nature *354,* 149–52.

Smiley, J.A. and Benkovic, S.J. (1994) Selection of catalytic antibodies for a biosynthetic reaction from a combinatorial cDNA library by complementation of an auxotrophic Escherichia coli: antibodies for orotate decarboxylation, Proc Natl Acad Sci USA *91*, 8319–23.
Smith, G.P. (1985) Filamentous fusion phage: novel expression vectors that display cloned antigens on the virion surface, Science *228*, 1315–7.
Smith, G.P. (1988) Filamentous phages as cloning vectors, Biotechnology *10*, 61–83.
Smith, G.P. and Scott, J.K. (1993) Libraries of peptides and proteins displayed on filamentous phage, Methods Enzymol *217*, 228–57.
Soumillion, P., Jespers, L., Bouchet, M., Marchand-Brynaert, J., Winter, G. and Fastrez, J. (1994) Selection of beta-lactamase on filamentous bacteriophage by catalytic activity, J Mol Biol *237*, 415–22.
Stemmer, W.P. (1994) Rapid evolution of a protein *in vitro* by DNA shuffling, Nature *370*, 389–91.
Szostak, J.W. (1992) *in vitro* genetics, Trends Biochem Sci *17*, 89–93.
Tuerk, C. and Gold, L. (1990) Systematic evolution of ligands by exponential enrichment: RNA ligands to bacteriophage T4 DNA polymerase, Science *249*, 505–10.
Wang, C.- I., Yang, Q. and Craik, C.S. (1995) Isolation of a high affinity inhibitor of uPA by phage display of ecotin, J. Biol. Chem. *270*, 12250–6.
Wells, J.A. and Estell, D.A. (1988) Subtilisin – an enzyme designed to be engineered, Trends Biochem Sci *13*, 291–7.
Wells, J.A., et al., (1987) Recruitment of substrate-specificity properties from one enzyme into a related one by protein engineering, Proc Natl Acad Sci USA *84*, 5167–71.
Wells, J.A., Vasser, M. and Powers, D.B. (1985) Cassette mutagenesis: an efficient method for generation of multiple mutations at defined sites, Gene *34*, 315–23.
Wilks, H.M., et al., (1988) A specific, highly active malate dehydrogenase by redesign of a lactate dehydrogenase framework, Science *242*, 1541–4.
Winter, G., Griffiths, A.D., Hawkins, R.E. and Hoogenboom, H.R. (1994) Making antibodies by phage display technology, Annu Rev Immunol *12*, 433–55.

Authors

Jon Appel
Torrey Pines Institute for
Molecular Studies
3550 General Atomics Court
San Diego, CA 92121, USA

Dr. *Elisabetta Bianchi*
Synthetic Peptides Unit
IRBM
Via Pontina km. 30.600
00040 Pomezia, RM, Italy

Sylvie E. Blondelle
Torrey Pines Institute for
Molecular Studies
3550 General Atomics Court
San Diego, CA 92121, USA

Dr. *Luisa Castagnoli*
Università di Roma "Tor Vergata"
Dipartimento di Biologia
Via E. Carnevale
00173 Roma, Italy

Giovanny Cesareni
Professor of Molecular Genetics
Università di Roma "Tor Vergata"
Dipartimento di Biologia
Via E. Carnevale
00173 Roma, Italy

Dr. *David J. Chiswell*
General Manager
Cambridge Antibody Technology Limited
The Science Park
Melbourn, Cambridgeshire, SG8 6EJ, UK

Dr. *Manuela Helmer Citterich*
Università di Roma "Tor Vergata"
Dipartimento di Biologia
Via E. Carnevale
00173 Roma, Italy

Riccardo Cortese, M.D., Ph.D.
Scientific Director
IRBM
Via Pontina km. 30,600
00040 Pomezia, RM, Italy

Charles S. Craik, Ph.D.
Professor of Biochemistry/
Biophysics and Pharmaceutical Chemistry
University of California, San Francisco
Department of Pharmaceutical Chemistry
513 Parnassus Avenue
S. Francisco, CA 94143-0446, USA

Dr. Luciana Dente
Università di Roma "Tor Vergata"
Dipartimento di Biologia
Via E. Carnevale
00173 Roma, Italy

Colette C. Dooley
Torrey Pines Institute for
Molecular Studies
3550 General Atomics Court
San Diego, CA 92121, USA

Barbara Dörner
Torrey Pines Institute for
Molecular Studies
3550 General Atomics Court
San Diego, CA 92121, USA

Dr. *Laura V. Doyle*
Chiron Corporation
Department of Chemical Therapeutics
4560 Horton Street
Emeryville, CA 94608-2916, USA

Dr. *Michael V. Doyle*
Chiron Corporation
Department of Chemical Therapeutics
4560 Horton Street
Emeryville, CA 94608-2916, USA

Dr. *Robert Drummond*
Chiron Corporation
Department of Chemical Therapeutics
4560 Horton Street
Emeryville, CA 94608-2916, USA

Dr. *John Earnshaw*
Cambridge Antibody Technology Limited
The Science Park
Melbourn, Cambridgeshire, SG8 6EJ, UK

Jutta Eichler
Torrey Pines Institute for
Molecular Studies
3550 General Atomics Court
San Diego, CA 92121, USA

Dr. *Franco Felici*
Biotechnology Department
IRBM
Via Pontina km. 30,600
00040 Pomezia, RM, Italy

Kevin FitzGerald
Cambridge Antibody Technology Limited
The Science Park
Melbourn, Cambridgeshire, SG8 6EJ, UK

Dr. *Susan Fong*
Chiron Corporation
Department of Chemical Therapeutics
4560 Horton Street
Emeryville, CA 94608-2916, USA

Dr. *Giovanni Galfré*
Group Leader
Immunology Unit
IRBM
Via Pontina km. 30,600
00040 Pomezia, RM, Italy

Dr. *Robert J. Goodson*
Chiron Corporation
Department of Chemical Therapeutics
4560 Horton Street
Emeryville, CA 94608-2916, USA

Dr. *Linda C. Griffin*
Gilead Sciences Inc.
353 Lakeside Drive
Foster City, CA 94404, USA

Dr. *Jürgen Hammer*
Roche Milano Ricerche
Via Olgettina 58
20132 Milano, Italy

Richard A. Houghten, Ph.D.
Torrey Pines Institute for
Molecular Studies
3550 General Atomics Court
San Diego, CA 92121, USA

Dr. *Gioacchino Iannolo*
Università di Roma "Tor Vergata"
Dipartimento di Biologia
Via E. Carnevale
00173 Roma, Italy

Dr. *John Kenten*
Cambridge Antibody Technology Limited
The Science Park
Melbourn, Cambridgeshire, SG8 6EJ, UK

Petr Kocis
Selectide
Dept. of Chemistry
Selectide Corporation
1580 E. Hanley Boulevard
Tucson, AZ 85737, USA

Achim Kramer
Universitätsklinikum Charité
Medizinische Fakultät der
Humboldt-Universität zu Berlin
Institut für Medizinische Immunologie
Schumannstr. 20/21
10098 Berlin, Germany

Viktor Krchňák
Selectide
Dept. of Chemistry
Selectide Corporation
1580 E. Hanley Boulevard
Tucson, AZ 85737, USA

Kit S. Lam
Selectide
Dept. of Chemistry
Selectide Corporation
1580 E. Hanley Boulevard
Tucson, AZ 85737, USA

Dr. *Alessandra Lanfrancotti*
Università di Roma "Tor Vergata"
Dipartimento di Biologia
Via E. Carnevale
00173 Roma, Italy

Michal Lebl
Selectide
Dept. of Chemistry
Selectide Corporation
1580 E. Hanley Boulevard
Tucson, AZ 85737, USA

Lawrence L.K. Leung, M.D.
Director, Vascular Biology and Medicine
Gilead Sciences Inc.
353 Lakeside Drive
Foster City, CA 94404, USA

Jon R. Lorsch
Harvard Medical School
Department of Molecular Biology
Massachusetts General Hospital
Boston, MA 02114, USA

Dr. *Alessandra Luzzago*
Biotechnology Department
IRBM
Via Pontina km. 30,600
00040 Pomezia, RM, Italy

Dr. *John McCafferty*
Cambridge Antibody Technology Limited
The Science Park
Melbourn, Cambridgeshire, SG8 6EJ, UK

Monica Mecchia
Biotechnology Department
IRBM
Via Pontina km. 30,600
00040 Pomezia, RM, Italy

Carmela Mennuni
Biotechnology Department
IRBM
Via Pontina km. 30,600
00040 Pomezia, RM, Italy

Dr. *Annalisa Meola*
Immunology Unit
IRBM
Via Pontina km. 30,600
00040 Pomezia, RM, Italy

Dr. *Olga Minenkova*
Università di Roma "Tor Vergata"
Dipartimento di Biologia
Via E. Carnevale
00173 Roma, Italy

Dr. *Paolo Monaci*
Biotechnology Department
IRBM
Via Pontina km. 30,600
00040 Pomezia, RM, Italy

Dr. *Alfredo Nicosia*
Biotechnology Department
IRBM
Via Pontina km. 30,600
00040 Pomezia, RM, Italy

John M. Ostresh
Torrey Pines Institute for
Molecular Studies
3550 General Atomics Court
San Diego, CA 92121, USA

Dr. *Lootsee Panganiban*
Chiron Corporation
Department of Chemical Therapeutics
4560 Horton Street
Emeryville, CA 94608-2916, USA

Marcel Patek
Selectide
Dept. of Chemistry
Selectide Corporation
1580 E. Hanley Boulevard
Tucson, AZ 85737, USA

Enrique Pérez-Payà
Torrey Pines Institute for
Molecular Studies
3550 General Atomics Court
San Diego, CA 92121, USA

Dr. *Antonello Pessi*
Head of Synthetic Peptides Unit
IRBM
Via Pontina km. 30,600
00040 Pomezia, RM, Italy

Clemencia Pinilla
Torrey Pines Institute for
Molecular Studies
3550 General Atomics Court
San Diego, CA 92121, USA

Caterina Prezzi
Biotechnology Department
IRBM
Via Pontina km. 30,600
00040 Pomezia, RM, Italy

Ulrich Reineke
Universitätsklinikum Charité
Medizinische Fakultät der
Humboldt-Universität zu Berlin
Institut für Medizinische Immunologie
Schumannstr. 20/21
10098 Berlin, Germany

Dr. *Steven Rosenberg*
Chiron Corporation
Department of Chemical Therapeutics
4560 Horton Street
Emeryville, CA 94608-2916, USA

Dr. *Jens Schneider-Mergener*
Universitätsklinikum Charité
Medizinische Fakultät der
Humboldt-Universität zu Berlin
Institut für Medizinische Immunologie
Schumannstr. 20/21
10098 Berlin, Germany

Nicolai F. Sepetov
Selectide
Dept. of Chemistry
Selectide Corporation
1580 E. Hanley Boulevard
Tucson, AZ 85737, USA

Francesco Sinigaglia, M.D.
Roche Milano Ricerche
Via Olgettina, 58
20132 Milano, Italy

Dr. *Rodger Smith*
Cambridge Antibody Technology Limited
The Science Park
Melbourn, Cambridgeshire, SG8 6EJ, UK

Maurizio Sollazzo, Ph.D.
Group Leader
Biotechnology Department
IRBM
Via Pontina km. 30,600
00040 Pomezia, RM, Italy

Jack W. Szostak, Ph.D.
Professor of Genetics
Harvard Medical School
Department of Molecular Biology
Massachusetts General Hospital
Boston, MA 02114, USA

Dr. *Costantino Vetriani*
Università di Roma "Tor Vergata"
Dipartimento di Biologia
Via E. Carnevale
00173 Roma, Italy

Cheng-I Wang
Biophysics and Pharmaceutical Chemistry
University of California, San Francisco
Department of Pharmaceutical Chemistry
513 Parnassus Avenue
San Francisco, CA 94143-0446, USA

Dr. *Richard Williams*
Cambridge Antibody Technology Limited
The Science Park
Melbourn, Cambridgeshire, SG8 6EJ, UK

Dr. *Jill Winter*
Chiron Corporation
Department of Chemical Therapeutics
4560 Horton Street
Emeryville, CA 94608-2916, USA

Qing Yang
Biophysics and Pharmaceutical Chemistry
University of California, San Francisco
Department of Pharmaceutical Chemistry
513 Parnassus Avenue
San Francisco, CA 94143-0446, USA

Dr. *Adriana Zucconi*
Università di Roma "Tor Vergata"
Dipartimento di Biologia
Via E. Carnevale
00173 Roma, Italy

Index